Organic Food Production in India

Status, Strategy and Scope

Organic Food Production in India
Status, Strategy and Scope

Chandralekha Arora

Oxford Book Company
Jaipur ▪ New Delhi

Published By

Oxford Book Company

Regd. Off. "Bony Residency", Gate No. 2
Opp. Tilak Public School, Vishweswariya Nagar,
Gopalpura Road, Jaipur - 302018, India
Ph: 0141 - 2761280, Telefax: 2761381
email - oxfordbookjpr@gmail.com
website - oxfordbookcompany.com

Delhi Office: 102, 1st Floor, Satyam House,
4327/3, Ansari Road, Darya Ganj,
New Delhi - 110002, India
Ph: 011-45652440

Distributed By
KEYA INTERNATIONAL PRESS LLP
E-mail: keyainternationalpressllp@gmail.com

Published in India, 2024

ISBN : 9788189473785

Price: ₹ 1900.00

Typeset by Shivangi Computers, Jaipur

Printed and bound in India by Thomson Press Ltd.,
New Delhi

Preface

The interest in health foods in India has grown rapidly in the past decade spurred by such factors as the desire for wholesome food and the anxiety caused by studies linking certain food additives with cancer and other diseases. As a result the incentive to practice organic farming has increased multifold, leading to a whole new category of agriculturalist products. Divested of the use of chemical fertilisers or pesticides, organic farming is a sub-activity of sustainable agriculture, striving to attain nutritional superiority of food with eco-friendly farm practice.

This text has been written in order to fulfil the need for readers to know all about organic farming. The emphasis is on acquainting readers with what entails organic food production, what it projects for the Indian agricultural scenario, the challenges and hurdles encountered in propagating it, its advantages and future. Emphasis has been put upon delineating the factors which are essential to its successful practice, strategies involved in implementing organic farming, and its current status in the country. In addition critical perspectives from eminent agronomists have been included in the book so as to foster greater understanding of the subject.

Chandrelekha Arora

Contents

1

Introduction

In 2005, only 1% of the Indian population actually consumed the organic food produced in their own country, primarily because the products were too expensive for the average person, costing as much as 30% more than the same non-organic food. High prices were mainly due to the process of becoming a certified organic farm. This can be very costly as years must pass for intensively farmed land to convert to organic.

The fact that more human labour is required in order to provide soil with the care of cultivation it needs to produce quality organic produce, also played a part. Of course, the fact that the rest of the world is willing to pay these high prices for the benefits of organically grown merchandise also plays a role in setting the average price paid by Indian consumers – the price charged by mainstream, commercial outlets that is. However, as organic farming in India has expanded and become established, the costs have been coming down and many more Indian people are now, thankfully, able to afford them.

There is more and more evidence that chemical based fertilisers, herbicides and pesticides are extremely harmful to our health. Conventional agriculture relies heavily on chemical fertilisers and toxic pesticides, which are harmful to human health, and which enter the food supply, penetrate water sources, harm the livestock, deplete the soil, devastate natural ecosystems, and contaminate the foods grown for the general public. In addition to which, there are over 7000 artificial and/or chemical preservatives and additives permitted in non-organic food including colourings, stabilisers, fillers, residual antibiotics, hydrogenated fats, etc. none of which are permitted in organic foods.

Not only do whole communities become subject to the dangers of toxic agricultural chemicals, they seep into the food on which they are applied; and the foods grown for the general pubic become contaminated as well. In many places trees have stopped bearing fruit because the heavy use of pesticides has killed the pollinators – the bees and butterflies. In India as elsewhere, chemical fertilisers and pesticides have been aggressively promoted and heavily subsidised to keep the fertiliser companies afloat, without realising the resulting devastation these chemical applications have wrought on the sustainability of agriculture.

The most recent introduction of GM seeds and crops, which need to be purchased every year at a high cost, and need to be heavily sprayed with the killer pesticides that the same companies produce, evokes a strong response from health and environment conscious people all over the world. Many of India's farmers have lost much of their ancestral farming knowledge and ancient seed varieties since the introduction of chemical based commercial agriculture and hybrid seeds.

In many areas of India and all over the world, valuable medicinal plants are becoming endangered because of indiscriminate commercial over-harvesting and other adverse collection practices that damage essential natural habitats. Proper organic agricultural practices require training and supervision for sustainable procurement methods that preserve rare and endangered wild plant species and the earth's natural biodiversity. Organic agriculture requires the elimination of commercial fertilisers, herbicides, pesticides and other dangerous chemicals. This in turn affects the whole community and creates a significantly healthier working and living environment for the farmers, their families and livestock, as well as the local wildlife. Through organic agriculture, farmers learn healthy, sustainable farming practices, and are regaining the wisdom of their agricultural heritage.

Organic agriculture enriches the soil season by season. It increases the long-term yield, nutrient value and potency of their crops. It allows for a naturally clean water supply, and provides overall richness, health and well-being of their families, livestock, farmlands and communities.

GROWTH OF ORGANIC FARMING IN INDIA

If you played the inspirational video above, you will have become aware of the company, Organic India. In the mid to late 1990's, it was the founders of this company that introduced organic farming methods to a small group of Indian farmers. These farmers, along with many others, had been watching the health of their land decline through the use of toxic, synthetic chemical pesticides, fertilisers, herbicides and fungicides.

The wildlife adjacent to their farms was also suffering as natural habitats were destroyed by conventional, industrial farming techniques for obtaining more crops, more quickly. As a result of these first steps taken by Organic India and the farmers with whom it worked, organic farming took off in India. Revenue generated by exports are bringing a better quality of life to organic farmers, their workers and families. Farming organically is bringing fantastic benefits in health to the environment and to those who consume the foods it produces.

Now, there are at least 15,000 certified organic farms in India and that number is growing rapidly. There are just as many small farmers using the same organic practices to grow their foods. As the domestic demand for organic food in India continues to grow, more stores and markets are carrying both certified and uncertified foods that claim to be organic. While organic exports must be certified, there are no laws in effect as of yet to control natural foods sold domestically.

A lack of knowledge among the general population and the absence of an organised system to regulate the sale of organic products create a great deal of confusion regarding the difference between natural and organic foods sold within India. However, organic markets are becoming increasingly common throughout the country and this helps community members find foods that are truly organic, being produced at the many known and trusted small local farms that do not seek certification for whatever reason.

The first organic food store in India opened in 1997, before the heightened interest in domestically grown organics. Greenway still supplies the residents of Mumbai with affordable organic products from about 20 local farms that use chemical free, organic cultivation. Conscious Foods was started two years later, also supplied by small local farmers, and today there are organic grocers, organic markets, and even organic restaurants in every major city in India. The market for organic food in India has seen tremendous growth over the past decade.

India is quickly becoming an organic centre for all types of crops and is making terrific progress towards supplying their own residents with the high quality and healthy benefits of organic food. Organic farming is an age old practice in India which got disturbed during First Generation Green Revolution. Organic manure has been practiced in Indian agriculture since the Vedic period. Sir Albert Howard, a British Agronomist, however, started the organic agriculture way back in 1900.

In organic farming, farm is viewed as an economy. It is a way of farming which excludes the use of off-farm inputs such as chemical fertiliser, pesticides, insecticides etc and is primarily based on the principle: use of natural organic inputs and biological plant protection measures. Properly managed organic farming reduces or eliminates water pollution and helps conserve water and soil on the farm thereby enhances sustainability and agro-biodiversity.

Indian farmers have an inherent understanding of how to work closely with the nature, because only 30 per cent of India's total cultivable area is covered with fertilisers because of assured irrigation and the balance 70 per cent of arable land is mainly rain-fed with little or zero use of fertiliser. It isn't a profit oriented but social-profit oriented. Commitment to nature protection is a prerequisite to organic farming.

India has comparative advantage over many other countries because of the vast cultivated area, which has remained free of contamination from chemicals, spread over distinctly varying agro climatic conditions, for example, large area in northeast region, northern hills and rain fed regions with very low or zero use of agro chemicals, can be instantly converted to organic farming. Farmers in these areas often use organic manure as a source of nutrients that are readily available either in their own farm or in their locality. Organic farming is based upon an integrated relationship among soil, minerals, water, plants, micro flora, insects, animals and human beings. It creates productive landscapes and successfully reconciles food production and environmental conservation.

Organic management relies on local human resources and knowledge to enhance natural resource processes, respecting ecological carrying capacities. By reducing dependence on off-farm inputs and creating more balanced nutrient and energy flows, ecosystem resilience is strengthened, food security is increased and additional income are generated. Organic farming responds positively to all sustainable agriculture and rural development objectives and helps in maintaining soil fertility to improve crop production and socioeconomic conditions of the farmers. One of the biggest rewards of organic farming is healthy soil that is alive with beneficial organisms. These healthy microbes, fungi and bacteria keep the harmful bacteria and fungi that cause diseases in check.

Organic farmers, working with nature, build soil that protects their crops from diseases. They also try to be diligent about crop rotation. They do not plant the same crop in the same location repeatedly, which encourages the build-up of diseases and pests that plague that particular crop. Excessive application of pesticides and fertilisers has caused damage to the soil and environment besides affecting crop production. The consumption of pesticides and fertiliser over the period of time has increased substantially.

Pesticides residue is the second largest agent causing cancer, next to cigarettes. Besides, the pesticides and fertilisers residuals that persist in the soil are harmful to the beneficial soil micro organism and earthworms thereby resulting in degradation of soil fertility. Fertilisers have a short term positive effect on productivity but a long term negative effect on the environment where they remain for years after leaching and running off, contaminating ground water and water bodies. In the name of growing more to feed the earth, we have taken the wrong path to sustainability.

Farmers committing suicide in growing numbers every passing year. Another negative effect of this trend has been on the fortunes of the farming communities' around the globe. Despite the so-called increase in productivity, farmers practically every where around the globe have seen downtimes in their fortunes. Pesticides residue controversies in the recent past in the bottled drinking water as well in the aerated beverages in India hardly come as a surprise.

Pesticides, a non-bio-degradable but highly toxic, find their ways into ground water and water bodies contaminating them and rendering them unfit for human consumption. Pesticides go into the water in the first place only because of the practiced agricultural system. The pesticides problem compounded in India because many pesticides banned abroad are manufactured/ dumped and sold freely in India.

Organic in agriculture is a labelling term that denotes products have been produced in accordance with certain standards during food production, handling, processing and marketing stages, and certified by a duly constituted certification authority or body. The certification of organic farms is required to satisfy the consumers that the product is totally organic. Certification agency conducts the inspection that minimum requirements prescribed for organic agriculture is fully met and issues certificates.

The formation of the International Federation of Organic Agriculture Movement (IFOAM) in 1972 gave an international framework for the discussion and codification of internationally recognised principles of organic farming. FAO-WHO has officially declared that International guidelines on organically produced food products should be considered important for consumer protection and information as they facilitate trade. The Codex Alimentarious Commission, a joint FAO-WHO food standards programme, body that sets international food standards, started to develop guidelines for the production, processing, labelling and marketing of organically produced food in 1991.

The Codex guidelines are important for equivalent judgments under the rule of World Trade Organisation (WTO). The Codex Alimentarious Commission defines organic agriculture as a holistic food production management system, which promotes and enhances agro ecosystem health including biodiversity, biological cycles and soil biological activity. It emphasises the use of management practices in preference to the use of farm inputs, taking into account that regional conditions require locally adapted systems. But the commercial organic farming in India is still at a nascent stage.

According to IFOAM - SOEL (Stiftung Oekologie & Landbau) survey of February 2005, India has about 76,326 hectare of land under organic management managed by about 5147 certified organic farms, which is only 0.05 per cent of total agricultural land. According to Agricultural and Processed Food Products Export Development Authority (APEDA), a nodal agency involved in promoting Indian organic agriculture, about 67292 tons of organic products worth of Rs 7123 lakhs are being exported from India. According to APEDA, India has 2508 thousand of hectare area under organic farming including herbs collections from the forest area of Madhya Pradesh and Uttar Pradesh producing 119656 tons of organic products, 165700 numbers of seedlings and cuttings and 264000 litres of effective micro organism.

The Indian organic farming industry is estimated at US $ 20 millions and is almost entirely export oriented. But there are a number of farms in India which have either never been chemically managed/ cultivated or has converted back to organic farming because of their farmers' beliefs or purely for reason of economics. These thousands of farmers cultivating hundreds of thousands of acres of land are not classified as organic though they are. Their produce either sells in the open market along with conventionally grown produce at the same price or sells purely on goodwill and trust as organic through selected outlets and regular specialist bazaars.

These farmers may never opt for certification because of the costs involved as well as the extensive documentation that is required by certifiers. Organic farming yields other benefits in which the external costs of organic farming-erosion, chemical pollution to drinking water and death of birds are one third of the conventional farming. There is another argument which is relevant to India. Because organic farming does not depend on extensive inputs, it might help shift the balance towards smaller farmers.

The organically produced food items are superior in quality aspects as compared to synthetic chemicals. They are good in taste, flavour, and essential nutrients and at the same time, they are free from harmful or toxic chemicals. The use of pesticides has led to high levels of chemicals in the soil, water, air and even our bodies. Fertilisers have a short term effect on productivity but a long term negative effect on the environment contaminating water bodies. In fact organic farming is another revolution emerging as Viable Alternative of Farming. It has the potential to transform agriculture as the main tool for nature conservation.

Large scale shift to organic farming would bring better results and not reduce the supply of food gains which is the apprehension among few scientists. Indian farmers and the traders can exploit the huge demand, both the domestic as well as global market for organic products. To promote and develop the Indian organic movement and to maintain its quality system at par with the international standard as to receive reciprocal approval from other countries, a National Programme of Organic Production (NPOP) has been launched which provides an institutional mechanism for the improvement of National Standards for organic products through a National Accreditation Policy and Programmes.

In order to provide reorganisation to organic products destined for export to the US, APEDA on the behalf of Ministry of Commerce (MOC) had initiated a process of equivalence of Indian Standard under NPOP with US standards since May 2002. Recently, USAD has agreed to recognise our accreditation system and the certificate of organic agricultural product by certification bodies accredited under NPOP for import of organic product into US with effect from February 16, 2006.

USAD has accepted affixing the Indian Organic seal along with USAD organic seal on Indian products as well as certificate. This will help the exporter by reducing the cost of certification which they have to pay to the international agencies. Indian Organic seal is thus going to gain popularity. Indian farmers and traders can exploit the huge domestic as well as international market for organic products because eating organic foods help reduce the risk of heart attacks, strokes and cancer.

Definition of Organic Agriculture

"Organic agriculture is a system of farm design and management that creates an ecosystem which can achieve sustainable productivity without the use of artificial external inputs such as chemical fertilisers and pesticides."

The Indian Government's strategy for organic agriculture is covered by the National Programme for Organic Production which aims to promote sustainable production, environmental conservation, reduction in the use and import of agrochemicals, the promotion of export and rural development. This strategy is promoted by the Ministry of Agriculture, the Ministry of Environment and the Ministry of Commerce.

Support to organic agriculture

— *Economic*: A programme of direct economic support is provided to farmers; however, this is limited in scale. As an incentive to adopt organic farming Rs 10,000 per ha is provided both during the conversion period (3 years) and after. The total cost to the country will be Rs 3.5 million per year until 2006-2007. This financial support aims to compensate for losses, promote organic agriculture, support infrastructure development, for conducting feasibility studies and preparation of guidelines for organic production. Some other specific financial incentives do exist, such as tax reductions and preferential conditions to credit.

— *Research*: Although support to agricultural research is not specific to organic agriculture, backing is given to, for example, the development of biofertilisers and biological control of pests and diseases. The scheme developing biofertilisers has an outlay of US$1.3 million during the years 1998 to 2002, a regular form of extension, technical assistance and training for farmers. A programme also exists for the promotion of organic farms and gardens for self sufficiency at the community level. So far 4 model farms have been developed and replicated. There are now 10 farms following the model developed for the production of organic joha rice and 20 farms using the model for organic sugarcane production in Siphajar, near Guwahati. There are 10 farms producing organic pineapples in Jumerdhepa near Agartala using another model and a further 20 farms using the fourth model for organic passion fruit production in Mao, near Manipur.

— *Marketing*: There is no regular support for the marketing or certification of organic products, however, occasionally on a case by case basis, financial help can be made available.

International certification organisations

India has no locally based certification bodies, however, two international certification organisations have branches in India, namely IMO India, part of the Institute of

Marketecology, Switzerland and Skal India, subordinate to Skal International, from the Netherlands. ECOCERT International is also active within India, but does not have a local branch. These market for export to the European Union under article 11 of EU Regulation 2092/91, but also certify to standards of non-European Union countries.

A National Accreditation Policy has recently been approved in India for the accreditation of certification bodies. The certification bodies already working within India are now being accredited. The Norms that apply follow the standards of the FAO/WHO Codex Alimentarius Commission guidelines for the production, labelling and marketing of organically produced foods.

For the present the National Steering Committee for Organic Farming, set up under the Chairmanship of the Secretary of Commerce to the Government of India, together with members from the Ministry of Agriculture, is the competent authority which consents to such accreditation. However, the Ministry of Agriculture will be the nodal Ministry to develop the mechanism for organic standards in the course of time.

Organic market

In 2000-2001 there were approximately 41 000 ha of land certified as organic in India. At present, organic products sold on the domestic market receive a premium of about 20-30 percent over conventional products. Such organic food is usually sold directly from the farmer or through specialised shops and restaurants. As India has no national legislation for organic agriculture, there is no mandate stating that organic products labelled as such need to be certified, however, a national Indian Organic Logo will eventually be developed for export.

Many organic farmer are organised into farmers associations such as the Organic Tea Manufacturers Association, from Calcutta; the Delhi Trust for

Development, based in New Delhi and Vidharbha Organic Farming Association, from Hingarh Ghat. Other farmers associations exist but still remain to become active. India does not import any organic products, however the export market was estimated to be worth approximately US$700 000 in 2000-2001. The main market for exported products was the European Union and involved rice, pulses, sugarcane products, walnuts, fruit pulp, tea, coffee and spices.

Future developments

Large tracks of land in India are already cultivated with very low use of agrochemicals, and traditional production systems adopted by farmers for many perennial crops have been, by and large, without any use of fertilisers and pesticides. Although yields in these areas are low and production is principally for subsistence, there exists great scope for organic agriculture.

Work carried out in different universities and institutions suggests that the productivity in these areas could be improved further by adoption of organic techniques, for example, organic manure and biopesticides. Recently a package of organic practices for a number of crops has been standardised and is now at the implementation stage. There is growing appreciation for organically grown food, especially as it provides additional value to production. However, there are a number of challenges that India and its organic community need to face:

- awareness needs to be raised amongst producers, processors and consumers regarding organic agriculture and the potential on the domestic and export markets for organic products;
- domestic markets need to be developed and supported, and the role of NGOs encouraged;

— a holistic approach to organic farming needs to be encouraged, both with farmers but also at research institutes and universities;
— local certification should be developed and a database on organic farms and marketing of produce should be initiated and maintained.

ORGANIC CROP PRODUCTION IN INDIA

Indian agriculture is progressing in all spheres to keep up with the ever-increasing population. In the Tenth Five-Year Plan (2002-07), the government envisaged at least 4% growth rate per annum, so that food security is assured and employment opportunities in agriculture are created. The National Agricultural Policy also aimed to strengthen the national economic growth through efficient marketing to accelerate foreign exports by establishing Agri-Export Zones for various crops in major crop-growing areas, where technical know-how and facilities for export will be provided. Therefore, there is a shift from sustainable to commercial agriculture.

Farmers can increase their income from crop production. Similarly, in several developed countries, demand for organic food and fibre, and by-products is increasing day-by-day. This change is mainly due to overuse or misuse of chemicals, particularly synthetic insecticides, fungicides, herbicides, fertilizers, plant growth regulators, etc. that resulted in undesirable side effects not only in the agro-ecosystems, but also on human health and life systems of beneficial fauna and microorganisms.

These recent trends in consumerism have opened a new vista for agricultural production on large scale by adopting contract farming, which can transform small farmers into viable commercial producers by monitoring quality, quantity and cost of crop production, and can link production with consumption. This change can help develop markets and bring about changes in the present conventional agriculture which is expected in India in near future. This note therefore discusses the present situation and future strategies of community cultivation.

Contract Farming in Organic Agriculture

The organic agriculture includes growing of crops by a set of guidelines that prohibit the use of synthetic products/ chemicals such as fertilizers, pesticides, plant growth promoters and livestock additives. Therefore, soil fertility and pest management is achieved through cropping patterns, manure, biofertilisers, cultural practices and biopesticides, including plant-derived products. At present, this system seems to be an ideal and valid solution to produce food and fibre without chemicals and to protect the environment.

Further, consumers are becoming conscious and critical about the quality of food and by-products that affect their health though the toxicity depends, to some extent, on the type of food consumed. India being one of the signatories for the World Trade Organisation, liberalisation, privatisation and globalisation has opened new grounds for agricultural marketing. The National Programme for Organic Production of the Commerce Ministry, and the State Governments are actively promoting investment in support of promising ventures in organic agriculture.

Facilities for export and certification of farm produce and by-products are being provided in the agri-export zones by the Agricultural Processed Food Products Export Development Agency (APEDA), National Bank for Agriculture and Rural Development (NABARD), National Boards for commodities like spices, tea, cashew nut, coffee, etc., International Federation for Organic Agriculture Movement, and Regional or State Marketing Federations. Additionally, commercial banks have initiated such banking for facilitating loans.

In fact, Uttarakhand, Madhya Pradesh and Kerala have been declared as 'organic states', and the State Bank of India has announced Rs 101 crore for contract farming in Maharashtra and Goa. Contract farming is an agreement between farmers and processors and/or marketing firms for the scientific production and supply of a specified agricultural product at a frequently and mutually predetermined price. Technical guidance on cultivation practices, harvesting, storage, etc. and quality inputs at wholesale rate are assured by the tripartite contract.

The main objective is to increase crop production, improve quality farm produce and possibly minimise cultivation cost. The farmer is therefore compelled to provide the produce in a specific quantity and quality determined by the processor. Although legal protection is possible to both parties, the success depends upon physical, social and cultural conditions because all terms and conditions prescribed in the agreement are to be fully respected by concerned parties, so that the project gives an impetus to scientific planning and implementation of integrated crop cultivation.

Considering the present socioeconomic status of Indian farmers, contract farming seems to be an ideal option because this system would have certain advantages over the present crop production and marketing systems, such as:

— Profit in produce sale is possible by capitalising the scientific research in post-harvest technologies.

— Indian agriculture per se is becoming commercial due to global demand for a variety of foods and fibre, and food products.

— Any crop can be cultivated on a large area to obtain produce of uniform quality by adopting appropriate technology. Crop production is also possible on small land-holdings through cooperative/ corporate farming to enhance productivity and avoid admixture or inferior quality produce.

— Technology transfer becomes easier due to large-scale adoption.

— Risk involved due to fluctuation in market price is minimised. This point is relevant to the present strategy of farm economics as the Minimum Support Price is generally declared at the end of crop season and it often remains ambiguous.

— Commercial and nationalised banks are coming forward to finance contract farming through soft loans and are revising prime lending rates.

— Additional income from intercrops is certain due to crop diversification. Consolidation of small and marginal lands can make farming economically viable, resulting in higher (>30%) net returns than traditional/ conventional farming systems.

Experiences

Contract farming is implemented in three ways:

(i) procurement contract, wherein sale and purchase conditions are specified,

(ii) partial contract, where the purchaser provides some inputs and purchases the farm produce at predetermined prices and

(iii) total contract, where the purchaser supplies and manages all inputs, but the farmer has to lease land and supply labour.

These contracts can be executed with individual farmers, a group of farmers or their cooperatives. In India, contract farming is being implemented successfully for food crops in Punjab, Haryana, Maharashtra, Tamil Nadu, Rajasthan and Karnataka by private companies. In the fibre sector, cotton farming has been executed by Super Spring Mills in

Tamil Nadu and Appachi Mills in Kerala and Karnataka Textile Mills Association and Gokak Forbes in Karnataka, with the help of seed suppliers, insurance companies, banks, and produce purchasing organisations.

The salient feature of these contracts is that farmers are free to sell cotton in the open market, if the price offered by the companies is not attractive enough. The involvement of the local self-help groups had created interest in the farming communities having small land-holdings. This participatory approach has resulted in upgrading the livelihood of farmers, particularly through bank loans.

Perspectives

In the case of organic food produced by a large number of farmers, marketing has to be arranged at village or taluk level. At present, the Indian food market is valued at Rs 650,000 crore, but only 1.6% of the farm products is exported in comparison to USA (12%), the Netherlands, Canada (6%), Spain, UK, Italy, Brazil, Australia (3%), and Argentina, Denmark, Thailand (2%). New crop zones that have been created during the last few years may boost such exports, because export has advantages over local/retail marketing for the following reasons.

— Since the average growth for organic farming is about 20-25% and higher prices (15-50%) are possible in the international trade, export of organic food produced by contract farming will certainly increase in near future, as India is recognised as an international agricultural hub.

— The Government supports contract farming for all crops grown under organic agriculture and biotechnological opportunities are foreseen because of maintenance of soil health, avoidance of environmental pollution, no risk of chemical residues in food and fodder, opportunities for rural employment, etc. These parameters lead to sustainable agriculture.

— This system will augment the farmer's income; help in the conservation of natural resources, and will have positive effects on the socioeconomic aspects of the farming communities.

— The purchasing companies help farmers identify proper land for contract farming with organic cultivation, plan crop production, arrange timely supply of quality inputs, provide technical guidance throughout crop growth and harvesting, assure purchasing and provide opportunities to add value to the farm produce. Thus, marketing risks and uncertainty are eliminated.

Challenges

— Organic farming on contract basis requires a long-term approach as it is based on the trust of farmers, for which grassroot intervention is critical for its success.

— Although 44% of the country's GDP comes from retailing, retail marketing of organic food is limited compared to other countries, e.g. 85% in USA, 40% in Thailand, 35% in Brazil, 20% in China and only 2% in India.

— Convincing farmers about the economic benefits of contract farming through field demonstrations should be taken up urgently. This may consist of educating farmers about new farming techniques, marketing skills for organic food and fibre, seeking cooperation from business communities and firms, informing consumers about the ill-effects of chemicals, etc. Sufficient number of the trained staff for such extension work in villages is a prerequisite for the success of contract farming.

— Certification cost for organic produce is comparatively high and needs to be reduced drastically.

— Facilities for storing and verification of quality of agricultural commodities at village level are lacking.

— Participation of processing firms/ purchasers is limited, as they are not aware of the long-term benefits and they also need large areas, which is difficult in regions where small and fragmented land-holdings are common.
— Banks are ready to finance contract farming in organic agriculture to only those farmers who are not defaultters.
— Under unfavourable climatic conditions, the quality of farm produce cannot be maintained or the whole crop may fail. Crop insurance initiated by the Government needs to be implemented on a large scale. Also, future liabilities of each party in the event of unforeseen circumstances are not included in such contracts and the committee for redressal of disputes has not been set up at regional or state level.

REFERENCES

Anon., *The Hitavada*, Nagpur edn, 7 March 2005.

Basu, A. K. and Tanweer, A., *J. Indian Soc. Cotton Improv.*, 2005, 30, 1-20.

Mudholkar, N. J. and Gahukar, R. T., *Baliraja*, 2005, 36, 76–82.

2

Principles of Organic Food Production

Two main sources of general principles and requirements apply to organic agriculture at the international level. One is the Codex Alimentarius Guidelines for the Production, Processing, Labelling and Marketing of Organically Produced Foods: According to Codex, "Organic agriculture is a holistic production management system which promotes and enhances ecosystem health, including biological cycles and soil biological activity.

Organic agriculture is based on minimising the use of external inputs, avoiding the use of synthetic fertilisers and pesticides. Organic agriculture practices cannot ensure that products are completely free of residues, due to general environmental pollution. However, methods are used to minimise pollution of air, soil and water. Organic food handlers, processors and retailers adhere to standards to maintain the integrity of organic agriculture products.

The primary goal of organic agriculture is to optimise the health and productivity of interdependent communities of soil life, plants, animals and people." The primary goal of organic agriculture is to optimise the health and productivity of interdependent communities of soil life, plants, animals and people. The other is the International Federation of Organic Agriculture Movements (IFOAM), a private-sector international body, with some 750 member organisations in over 100 countries. IFOAM defines and regularly reviews, in consultation with its members, the Basic Standards that shape the "organic" term.

According to the IFOAM 2002 Basic Standards, "organic agriculture is a whole system approach based upon a set of processes resulting in a sustainable ecosystem, safe food, good nutrition, animal welfare and social justice. Organic production therefore is more than a system of production that includes or excludes certain inputs." In contrast to food labelled as "environmentally-friendly", "green" or "free-range", the label "organic" denotes compliance with specific production and processing methods.

Most synthetic pesticides and fertilisers, and all synthetic preservatives, genetically modified organisms, sewage sludge and irradiation are prohibited in all existing organic agriculture standards. Adherence to organic agriculture standards, including consumer protection against fraudulent practices, is ensured through inspection and certification. Most industrialised countries have regulations governing food labelled as "organic". Other terms also used are, depending on the language, "biological" or "ecological".

Organic agriculture principles are consonant with principles of biodynamic agriculture and permaculture. Started by Rudolf Steiner in 1924, biodynamic agriculture embraces holistic and spiritual understanding of nature and the farm within it, where the farm is a self-contained evolving organism which keeps external inputs to a minimum: biodynamic preparations are used and requirements include, among others, harmony of cultivation with cosmic rhythms, fair trade and the promotion of associative economic relations between producers, processors, traders and consumers.

Certification requirements of biodynamic agriculture (labelled under the Demeter International network in Africa, America, Australia and Europe) include many organic standards, which are recognised under the United Register of Organic Food Standards and governmental organic aid schemes. In the late 1970s, the ecologist Bill Mollison developed the concept of "permaculture" as an interdisciplinary earth science. Permaculture is a landscape and social design system that works to conserve energy on-farm or to generate more energy than it consumes.

Care for natural assemblies (including wilderness), rehabilitation of degraded land and local self-reliance are central to permaculture. Permaculture does not have distinct certification but this management approach is embraced by organic agriculture.

"Organic agriculture" is not limited to certified organic farms and products but includes all productive agricultural systems that use natural processes, rather than external inputs, to enhance agricultural productivity. Organic farmers adopt practices to conserve resources, enhance biodiversity, and maintain the ecosystem for sustainable production. This practice is often but not always oriented towards the market for food labelled as organic. Those who seek to label and market their foods as organic will usually seek certification—almost certainly if they grow to export.

However, many farmers practise organic techniques without seeking or receiving the premium price given to organic food in some markets. This includes many traditional farming systems found in developing countries. Traditional agriculture includes management practices that have evolved through centuries to create agricultural systems adapted to local environmental and cultural conditions. Owing to their nature, traditional systems do not use synthetic agricultural inputs. Many, but not all, traditional systems fully meet the production standards for organic agriculture.

"Organic Agriculture" is not Limited to certified organic farms and products but Includes all productive agriculture systems that use natural process, rather than external inputs to enhance agriculture productivity. It is important to distinguish certified from non-certified organic agriculture for the purpose of this study. Agriculture that meets organic production standards, but is not subject to organic inspection, certification and labelling is referred to as "non-certified organic agriculture" as distinguished from "certified organic agriculture." While economic and institutional conditions differ, both rely on the same technology and principles.

Although the results might be similar, non-certified organic agriculture may not always represent a deliberate choice between alternative production systems—lack of access to purchased inputs may constrain such choice. Whatever the motivation, an organic farm reflects an intentional management system in which a producer manages resources according to organic principles. Non-certified organic agriculture therefore includes traditional systems which do not use chemicals but which apply ecological approaches to enhance agricultural production.

By contrast, some agricultural systems simply do not use purchased inputs because the farmer cannot afford them nor has access to them. These systems cannot be considered as organic according to internationally recognised standards. These systems usually have low, declining productivity. Negligent systems often result in environmental degradation and can create public nuisances that threaten neighbouring farms as reservoirs for noxious weeds, pests and diseases.

Organic standards require operators to conserve, restore, and enhance natural processes; to work with nature to protect crops, rather than to submit to or subdue it. Producer's decision-making is therefore essential to the meaningful differentiation of organic agriculture from systems that do not use synthetic inputs by neglect. Production by neglect is not considered "organic" for the purpose of this study, even if the organic standards in some local jurisdictions have not made this distinction. All agricultural management systems that apply ecological approaches but which make use of some synthetic inputs and/or genetically modified organisms are obviously excluded from the organic category.

ORGANIC AGRICULTURE DEVELOPMENT

The organic agriculture sector is currently the fastest growing food sector. Growth rates in organic food sales have been in the range of 20-25 percent per year for over a decade. Growth rates of organic lands are impressive in Europe, Latin America and the United States. The total area of organic land tripled in Europe and the United States between 1995 and 2000. In the last 5 years, in Argentina, the organic land area increased 1 280 percent. However, these reported percentage increases must be viewed in the context of the low absolute levels.

Globally, certified organic agriculture occupies less than 1 percent of lands and 1-2 percent of food sales. In some cases, the growth may reflect the entry of land long farmed organically into a certification programme rather than an actual switch in farming systems. Prior to the late 1980s, the slow but constant development of organic agriculture was driven by grassroots organisations, farmers and traders. In the United States, the states of Oregon and California adopted organic legislation in 1974 and 1979, respectively.

In all other parts of the world, it took a long time before the standards established by the organic agriculture community were echoed by national and supranational legislation and control systems. The recognition of the role of organic agriculture in achieving environmental objectives, including sustainable use of land set aside, led to the adoption of agri-environmental measures to encourage organic agriculture. On the supply side, and in the European Union, policy instruments stimulated small farmers to convert to organic farming by providing financial compensation for losses incurred during conversion.

Consumers concerned with food quality, as well as the protection of the environment, were the first to stimulate demand. New market opportunities have developed as part of a business strategy to address consumer concerns, particularly in the European Union and the United States. Major food companies see the processing, handling, stocking, and promoting of organic foods as part of a positive public image. Retailers of all sizes now aggressively promote and market organic food, with major food retailing chains now accounting for a major share of the retail markets for fresh as well as processed foods.

Consumers are increasingly sceptical on the safety of conventional foods and the soundness of industrial agriculture. The use of growth regulators stimulated interest in organic food. The crisis over dioxin-contaminated food and livestock diseases further increased demand for organic food. Consumer surveys in almost every country show a segment that demands an alternative to genetically modified foods. Governments have responded to these concerns by setting targets for the expansion of organic production. Thus, the concern of consumers and governments with the quality and safety of food has become the major driving force in the development of organic agriculture in industrialised countries.

While some may question the validity of consumer concerns, there is no doubt that these have contributed to the growth of the organic sector. These concerns have also opened possible markets for developing country exporters, enabling them to enhance foreign exchange earnings and diversify their exports. Price premiums of between 10-50 percent over prices for non-organic products, as well as more secure markets for organic commodities, can help counterbalance the loss of preferential trade arrangements, falling food prices and withdrawal of government support to agricultural inputs and other services.

Major northern markets offer good prospects for suppliers of organic products not domestically produced. These include coffee, tea, cocoa, spices, sugar cane, tropical fruits and beverages, as well as fresh produce in the off-season. Increasingly, governments in developing countries are creating conditions in support of organic exports. Non-certified organic agriculture is of particular importance for meeting local food requirements while providing protection and sustainable use of natural resources.

Organic management makes it possible to save on production costs and to promote economic and/or food self-reliance. In market marginalised and resource-poor areas where farmers have no access to modern inputs and technologies, organic agriculture can also raise the productivity of traditional systems by optimizing the use of local resources. For example, hundreds of thousands of indigenous farmers have turned to the organic movement to reinstate along the Andes sophisticated agricultural practices developed by the Incas.

Individual small family vegetable plots and groups/associations managing organic produce for domestic urban markets and small informal fairs are widespread. Cuba has adopted organic agriculture as part of its official agricultural policy, with substantial investments in research and extension, to compensate for shortages in external input and to substitute food imports.

RESOURCE USE EFFICIENCY

Comparisons of the performance of organic and conventional agriculture systems are meaningful only when made over an intergenerational period of time in order to assess the continued capacity of natural resources to sustain agriculture. High yields in non-organic systems are often exploitative systems that degrade land, water, biodiversity and ecological services on which food production depends. Most comparisons of the efficiency of alternative production systems focus merely on the gross yield of marketable commodities.

Farmers usually experience a decline in yields after discarding synthetic inputs and converting their operations to organic production. After the agro-ecosystem is restored and organic management systems are fully implemented, yields increase significantly. If conversion to organic departs from low-input systems, then yields tend to be stable. A study from Kenya indicates that organic agriculture in the tropics shows a good performance. This finding was contrary to conventional wisdom that contended such a system would be constrained by insufficient organic material.

Organic systems in medium potential areas significantly outperformed conventional methods in maize grain yields, net cash benefits, return on capital and return per family labour day. For example, organically grown maize experienced less damage from weevils during storage than its conventional counterpart. This and other similar findings show that organic systems can double or triple the productivity of traditional systems.

The productivity of organic agriculture systems varies through the different stages of management:

(i) in-transition from conventional to organic management;

(ii) in-conversion from traditional to organic management;

(iii) organic management based on input substitution, and

(iv) complete shift to a systems approach.

The need to secure farm economic viability in the short-term results in few farms achieving a systems approach. In most cases, transitioning farmers suffer decreased productivity, but even then there are cases where the conventional system is in such decline that there are significant positive yield responses to the addition of organic matter, particularly when there is a carryover of synthetic N fertilizer such as calcium nitrate.

Also, there are cases where transitioning farmers still have some carryover with herbicides that suppresses weeds well into the second year of transition; for them the decline in yields related to weed pressure does not begin until the second or third year. This is most common in grain field crops such as wheat, rice and corn. Yield losses are greatly exaggerated by conventional experiment stations. The actual experience of most transitioning farmers is that the yield losses are tolerable. The increased labour seems to hurt US farmers more than lower yields.

Yield losses are caused by a number of interrelated factors: soil organic matter and biological activity take time to get established; many conventional farms are on a pesticide `treadmill' that does not permit the establishment of beneficial organisms for pest, weed, and disease suppression; and fertility problems are common until restoration of full biological activity. The degree of yield loss varies, and depends on inherent biological attributes of the farm, farmer expertise, the extent to which synthetic inputs were used under previous management, and the state of natural resource.

Sometimes a few years may suffice or it may take many years to restore the ecosystem to the point where organic production is viable in economic terms. There are some soils that are so marginal and highly depleted that it would not be possible to farm without substantial government subsidies for certain commodities. In the medium term, and as expertise increases, the value of organic agriculture becomes more evident because of yield improvements and increased fertility of the agricultural system. In the longer term, the performance of organic agriculture increases in parallel with improvements in ecosystem functions and management skills.

The performance of organic agriculture cannot be judged based on the comparison of a single-crop or a single year. Organic agriculture usually performs better if one considers total production of useful crops per area. The greatest constraints faced by transitioning farmers are the lack knowledge, information sources, and technical support. Greater government investment in appropriate research and extension services can help overcome these constraints. Extreme weather fluctuations present a growing threat to agriculture.

Organic systems appear to be more stable and resilient in response to climate disruption based on comparisons with their conventional counterparts under stress conditions such as severe drought and flooding. In 1993, conventional rice in Japan was nearly wiped out by an unusually cold summer while organic farmers yielded 60-80 percent of the annual average. The better composition of water-stable aggregates in organic soils and reduced soil compaction result in the favourable performance of organic systems under both flood and drought conditions.

In market marginalised and resource poor areas where farmers have no access to modern input and technologies, organic agriculture can raise the productivity of traditional system by optimising the use of local resources. Yield comparisons offer a limited, narrow, and often misleading picture of the different production systems. Profitability and long-term economic viability would be a better indicator to determine what techniques operators would choose. Organic production, processing and marketing depend on several interrelated factors.

Profitability in organic agriculture depends on relative input and labour costs, actual production costs, conventional and organic market conditions and the price premiums received for organic commodities. In some countries, government payments for the adoption of organic techniques are available to offset certain costs and encourage practices that have long-term benefits to the environment. Accurate comparisons of economic performance must take account of performance over a complete rotation rather than a single year. Moreover, the multiple environmental benefits of organic farming, difficult to quantify in monetary terms, are essential ingredients in any comparison.

Economic Performance

Only a few studies have assessed the long-term profitability of organic agriculture systems. While these studies vary in their methodologies and conclusions, they consistently show high revenues relative to conventional agriculture because of the premiums received. However, the costs relate to whole farm production over the whole rotation period: this includes both marketed products and non-food products.

In particular, incomes achieved in one season may appear high because of price premiums but low in the subsequent rotation seasons if these crops have low or no market values. Looking at these seasons individually does not give an accurate picture of the financial viability of organic agriculture. Unfortunately, many comparative studies of organic and conventional or integrated production focus on a single crop and a single year.

The economic performance of organic agriculture in Europe shows a situation where organic farmers receive financial support and premium prices but where labour is expensive. An extensive analysis of European farm economics in terms of labour use, yields, prices, costs and support payments concluded that profits on organic farms are, on average, comparable to those on conventional farms. In the United States, profits are also comparable, despite the fact that there are no direct subsidies for organic agriculture.

In developing countries where organic agriculture is not subsidised, synthetic inputs are expensive and labour is relatively cheap, market-oriented organic farmers can achieve higher returns thanks to reduced production costs and diversified production. For example, in the Philippines, price premiums are not a sufficient incentive to market rice as organic. Farmers have adopted organic practices nevertheless because the avoidance of external inputs saves on production costs while yields are more stable.

In Madagascar, hundreds of farmers have increased their irrigated rice yields from two to as much as eight tonnes per hectare by using local seeds, composts and innovative soil, plant, water and nutrient management practices. Organic products tend to command impressive premiums in developed countries at retail level: on average between 10-50 percent above conventional prices for the same commodity. These premiums reflect several underlying factors on both the demand and supply side. Premiums reflect strong consumer demand, with some consumers willing to pay higher premiums than others do.

Most consumers in developed countries will pay a premium for organic, but only to a point. As the premium increases, the number of consumers willing to pay it decreases, because the conventional commodity is always available as a substitute. Premiums in excess of 50 percent usually have underlying supply constraints and bottlenecks. These are often temporary and unpredictable. Because fewer fungicides and post-harvest tools are available, fresh organic produce tends to be more seasonal and local. Long-term high premiums often reflect severe production problems related to chronic endemic pests and diseases that cannot be managed effectively by existing biological and cultural techniques.

On the other hand, situations where organic production costs are as low as, or lower than, conventional production will frequently see little or no organic price premium received by farmers. Premiums compensate farmers for skilled resource management; higher labour costs, unit costs, and handling expenses; and administrative, inspection, and certification fees.

Premiums also reflect the price to avoid and mitigate negative environmental externalities incurred by conventional agriculture. These include costs of damage to natural capital, human health and reduction of water, air and soil pollution. Such indirect costs are not usually included in food prices, and this distorts the market while encouraging activities that are costly to society.

Many reasons contribute to the additional costs to market organic products: inspection and certification fees, segregated storage, fewer options to control post-harvest pests and diseases, properly cleaned and well-documented transportation, careful handling to avoid dilution and contamination, appropriate packaging, and economies of scale. Because organic producers comprise a smaller proportion of the agricultural industry, individual producers are widely dispersed.

Greater collection and assembly costs further add to the costs of transportation. Pest and disease infestations can result in handlers facing an unpleasant choice between losing most and possibly all of a crop, or treating it with a prohibited substance to recover the losses, and selling the product as non-organic. Segregation increases costs of handling. Many retailers require their suppliers to provide individual packaging and special labelling for organic food not required for conventional food.

To date, consumers in industrialised countries have been willing to pay a premium for organic food because they perceive environmental, health, or other benefits from that choice. While surveys show that consumer demand is unmet, organic farmers also report insufficient demand for their products. Balanced expansion of supply and demand and reduction in organic production costs will be one factor in maintaining the organic price premium.

At present, the general tendency for demand to outgrow supply suggests that the premium is not under immediate threat for most product categories. Organic production is expected to continue to offer premium prices and a profitable alternative to conventional production systems for many farmers. High prices and limited outlets have historically curtailed demand for organic agriculture. Lower prices would expand the organic market without discouraging producers, provided that the premium still compensates the costs of transition and provides a living wage to the producer.

Most of the premium is captured by retailers, wholesalers, distributors and processors. Retailers can reduce prices while maintaining the profitability of organic farmers. A price premium to producers of 10-20 percent, perhaps even 50 percent, would have almost no impact on consumers. This, however, is not likely to be accepted by retailers and where feasible, direct marketing channels are being developed. At present, the marketing strategy of many major food retail chains is to expand the supply of a few low-cost organic products produced by a relatively small number of producers. This strategy benefits a few large organic farms, which rely on input substitution and global market over small and medium-sized local farmers.

Greater investments in research and extension offer long-term solutions to organic production and handling constraints. The redirection of only a small portion of public expenditures towards biological and cultural methods could increase yields, lower handling, distribution, and marketing costs, and deliver lower prices for organic food to consumers. As the organic market grows and matures, economies of scale should narrow margins for conventional products.

In many developing countries, there are no domestic institutions that can assist farmers to produce, handle, and market organic food. Extension services deter the adoption of organic agriculture because agents are trained to advise farmers to use what the experiment station has determined to be the most efficient inputs. Most often, organic methods are considered obsolete throwbacks to a less efficient time. Land tenure is another determining factor, given the long-term commitment needed for organic methods to be effective.

Tenant farmers are unlikely to invest the necessary labour and sustain the costly conversion period without a guarantee of continued access to the land. Organic farmers may take years to reap a return on their investment and tenants and sharecroppers seldom have the luxury to wait that long. The trade of organic commodities needs to be viewed with reference to international commodity markets. Prices received by farmers for conventional products have stagnated or decreased in real terms over the past thirty years, with farmers sometimes collecting revenues below production costs.

Relative decline in prices since the late 1990s affects almost all agricultural commodities. Low coffee prices have forced millions of small farmers into crippling debt that ultimately results in them forfeiting their land. Low world prices also mean low export earnings: developing countries' export earnings from beverage crops fell by 18 percent between 1999-2000. Current low price levels of agricultural commodities are likely to remain so in the short-term. Costs of conventional farming inputs have increased substantially, usually requiring hard currency to import.

Organic agriculture offers an opportunity to improve income because of: consumers' willingness to pay price premiums for organic produce and lower production costs due to reduced use or absence of imported inputs. Even when the price premium on organic products is low, stable and profitable long-term prices offer more security to farmers than volatile conventional commodity markets.

Social Performance

The conversion of a farm to organic practices influences all facets of the operation, including labour demand, social structures, and decision-making processes. Organic agriculture enterprises often require more labour input than conventional farming in order to replace external energy and capital inputs such as fertilisers and herbicides. However, the extent depends on the intensity of the operation and level of farm capitalisation.

Crop diversification on organic farms, and related different planting and harvesting schedules, distributes labour demand through the season. This stabilises employment, reduces turnover, alleviates many problems related to migrant labour, and helps to spread the overhead payroll costs per employee. Diversity in agricultural production and value added products increase income-generating opportunities and spreads the risks of failure over a wider range of crops and products.

Organic Production

Labour demand is at the same time a constraint to organic conversion and an opportunity to expand employment in rural communities. Northern countries with high wages and a declining rural population have the greatest difficulties in finding an adequate labour supply. This constraint is however overcome by employing labour force migrating from developing countries and in-transition economy countries.

The introduction of organic agriculture may shift gender distribution of labour insofar as men may prefer to be involved with mechanised agriculture. In developing countries, women depend on access to common property because they seldom own land. Given that credit frequently requires land as collateral, landless people in general, and women in particular, find themselves unable to obtain credit through most lending institutions. It might therefore be assumed that organic agriculture facilitates women's participation as it does not rely on purchased inputs and thus reduces the need for credit.

However, since organic agriculture requires several years to improve the soil, insecure long-term access to the land is a major disincentive for both women and men for launching into such a long-term enterprise. To be competitive, organic operators need to experiment with new techniques, and must manage labour, land, and capital quite differently from conventional operators.

The different options presented to farmers result in a variety of choices in techniques. Some succeed, some fail, and the difference is often based on experiments undertaken by the farmers, whether collaboratively with, or independent of, the public research institutions. On-farm research generates new knowledge that is shared with other farmers. Such learning processes lead to greater innovation together with increased likelihood of these technologies to persist.

Organic Agriculture and Local Self-Reliance

By building on local knowledge, organic agriculture approaches revitalise traditional customs and local self-reliance. Employment opportunities and higher returns on labour encourage people to remain in agriculture, reinvigorating rural communities. Strengthened social cohesion and partnerships within the organic community make for better connections with external institutions.

Organised groups, such as producer cooperatives, have better access to markets and can negotiate their needs as equal partners in the food supply chain. Together with the production system, the social environment of those engaged in organic agriculture generally improves: in fact, many organic systems incorporate fair trade principles which improve working conditions. The IFOAM Basic Standards includes a chapter on Social Justice Standards. These refer to and are based upon the conventions of the International Labour Organisation on labour welfare and to the human rights charters of the United Nations.

The 2002 version of the IFOAM Basic Standards proposes to ensure access for workers, farm families and indigenous people to bargain collectively for fair wages, safe and healthy working conditions, and social services. A growing number of certified organic agriculture commodities produced by small-scale farmers organised in democratic cooperatives meet fair trade requirements: farmers are paid adequately to cover costs of production and a social premium to improve the quality of life.

Although the organic movement shares a consensus that social requirements are necessary, specific standards are controversial. Standard-setting bodies are sensitive to national sovereignty and the cultural context governing social and economic relations. Such standards might create trade barriers to some developing countries organic exports, but this pressure may trigger social and economic reform in many countries.

Institutional Performance

Because "organic" is a production process claim, consumers must rely on certification programmes that verify claims. The standards that specify the organic production process appear quite precise, especially compared with claims made for other competing products. Consumers decide to purchase organic food in part because that choice reflects their values. For example, many consumers demand that organic food meet strict animal welfare standards. Others expect organic to mean fresh, local and minimally processed.

The whole organic community requires that genetically modified organisms not be used to produce or process organic food or fibre. Social and economic issues, including producers receiving a fair price, are increasingly receiving attention. Most industrialised countries have regulations that govern organic agriculture, including Australia and the European Union countries, Switzerland, Japan and the United States.

Some developing countries have also established policies and regulations on organic agriculture. In February 2002, 56 countries were reported to be at some stage of regulating the organic sector. Differences with regards to the scope of regulations and variation in their implementation raise a number of concerns, namely:

- import discrimination whereby compliance is required with standards not always suitable to agro-ecological conditions of export countries;
- multiple accreditation of certification bodies in order to access the three main organic agriculture markets (Europe, Japan and the United States);
- difficulties to traders, due to different rules interpretation by certification bodies;
- enormous workload (and delays) for authorities to negotiate bilateral equivalency;
- limitation of bilateral agreements with regards to products with ingredients sourced from around the globe.

The Codex Alimentarius Guidelines for the Production, Processing, Labelling and Marketing of Organically Produced Foods constitute a recognised basis to harmonise organic standards and national regulations. The Codex Committee on Food Import and Export Inspection and Certification Systems is developing Draft Guidelines on the

Judgement of Equivalence of Technical Regulations Associated with Foods Inspection and Certification Systems which intend to develop an infrastructure for the review of technical requirements other than sanitary measures associated with inspection and certification systems. These Guidelines suggest a process and general principles to determine the equivalence of all food systems and are particularly relevant to organic agriculture.

In the absence of official measures to address equivalence, the organic community has organised an international programme to accredit certification bodies. The IFOAM Accreditation Programme, established in 1992, developed international procedures to evaluate organic certification programmes and assess compliance of organisations that claim to adhere to organic standards. Certification programmes involved with organic agriculture apply to the International Organic Accreditation Service (IOAS), an NGO established in 1997, in order to be accredited.

IOAS evaluates their standards against the IFOAM Basic Standards and examines the competence of their programme against established criteria through field visits and audits. To date, IOAS has accredited about 20 certification bodies, operating in both developed and developing countries. This privately organised service has the potential to facilitate international trade in organic products, but this promise will be fully realised only when it is recognised by governments which have developed rules for organic agriculture.

Developing countries are important suppliers of organic commodities. They, however, need to establish that they conform to the standards and rules of the importing developed countries. Suppliers to multiple markets may need to carry several such certifications: standards accepted in Sweden may not be recognised in the United States or Japan, and vice-versa. In countries where recognised domestic facilities are lacking, suppliers often must hire foreign inspection and certification bodies; in many instances, this is prohibitively expensive.

Given that a certain part of certification is a fixed cost, certification costs take a higher percentage of earnings of smaller units. Smallholders in developing countries have little chance to export certified organic products without active government support for inspection and certification. Alternative control systems for small holders are however developing in order to ensure quality assurance without depending totally on foreign inspectors and certification bodies. Many developing countries require external technical assistance to build the capacity for technical, organisational, and legal skills needed to establish reliable certification and accreditation programmes. Some certification bodies become accredited by the importing country. This requires educated, trained personnel and administrative structures.

International equivalence of various national organic standards will reduce the administrative overhead, improve public-sector relations with private certifiers and traders, and eliminate redundant certification. This bureaucracy and its attendant cost particularly burdens poor farmers in developing countries. Internationally recognised accreditation and equivalence will benefit exporting and importing countries alike because it ensures conformity with requirements of importers while recognising the competence and compliance of the exporters.

Ensuring compliance must be legitimate and enforced with sufficient sanctions in both developed and developing countries alike. There is evidence that fraud exists in many developing countries. Recent experiences in Germany and in the United States show that developed countries are not free of negligence and fraud. This hurts the honest legitimate organic farmer in the exporting country as well as the consumer in the importing country. Trusted and enforced organic guarantee systems are key to successful trading of organic agriculture products. The establishment of an internationally agreed accreditation mechanism is crucial for determining equivalence of imported organic products.

REFERENCES

Altieri, Miguel., "Sustainable agriculture," In *Encyclopedia of agricultural science,* v. 4. Charles J. Arntzen and Ellen M. Ritter, (eds.), San Diego, Academic Press, c1994. p. 239-247.

Horne, James E., and Maura McDermott., *The next green revolution: essential steps to a healthy, sustainable agriculture.* New York, Food Products Press, c2001, 312 p.

Klinkenborg, Verlyn., "A farming revolution: sustainable agriculture," *National geographic,* v. 188, Dec. 1995: 60-88.

Soule, Judith D., and Jon K[illegible]per., *Farming in nature's image: an ecological approach to agricultu[illegible].* Washington, Island Press, c1992. 286 p.

3

Organic Farming and Food Security

Food is accessible when individuals have access to adequate resources and entitlements for acquiring appropriate foods nutritious diets. Entitlements are defined as the set of all commodity bundles over which a person establish command given the legal, political, economics and social arrangements of the community which they live. If food insecurity is a knot of interrelated problems related to lack of access, then food security is a web of mutually supporting elements that strengthen productivity and independence over Improvements must occur on many levels simultaneously, yet in policy environments that are often unsupportive.

Given the complexity of this task, organic systems have a great potential to increase food access, reduce risk and build long-term investments that contribute to food security. A development approach focused on the means to produce or purchase food is urgently needed, especially for resource-poor and indigenous farmers, many of whom live in fragile environments, well as for rural populations whose livelihoods are increasingly challenged by macro-economic factors.

Access to food involves entitlements for producing or acquiring food, which are discussed this chapter under three main headings:

— Access to productive natural resources including land, water, agroforestry, biodiversity, seeds and other genetic resources, and environmental services: using organic agriculture, resources for productivity and long-term sustainability are managed simultaneously; these include resilience under environmental stress and climate change. Efficient use of local resources also more financially sustainable for farmers, reducing costs and improving net incomes.

— Access to diverse types of knowledge, including traditional and indigenous knowledge, current best practices for organic agriculture from other locations, locally relevant research and sitespecific observation: knowledge engages the power of human creativity, skills and experience towards more reliable access to food, leveraging all local assets towards the community's critical needs.

— Access to domestic and international markets in which fair prices and stable, long-contracts can be obtained: market opportunities are most often realised through the formation of effective farmer groups, which can leverage their efforts towards both improving local supplies and maximising returns from surpluses. Increasing production to meet expanding markets improves local labour markets and returns on labour costs. In addition, fair meaningful urban, non-farm employment opportunities are urgently needed to ensure access urban poor people to food entitlements.

Successful projects combine these elements to promote food security, self-sufficiency and selfreliance, through greater community control of agriculture and food systems. Impacts may be seen at the household, community, regional, and national levels. Ensuring access to food at the household and family levels means that there are adequate amounts of nutritious and locally appropriate food accessible to all regardless of gender, age or family status.

Ensuring access at the regional and national levels requires policy reforms along with significant financial and institutional support to rural infrastructure improvement, local capacitybuilding and empowerment of rural communities aimed at enhancing local food security.

While there are still only a limited number of global studies documenting the contributions of organic agriculture to increasing access to food, their findings provide a broad range of important insights, which highlight the urgent need for additional research and documentation of issues including:

— the role of organic agriculture in rural poverty alleviation;

— the role of non-certified organic agriculture in subsistence food systems; and

— the agro-ecology response to multinational corporations.

The International Fund for Agricultural Development (IFAD) has explored the link between organic agriculture and poverty reduction in Asia, South Asia and Latin America. Julia Johannsen and colleagues at the German NGO Forum on Environment and Development have discussed the flexibility of organic systems for reducing hunger in a variety of environmental and socio-economic situations. Jules Pretty and colleagues at the University of Essex have described how organic agriculture makes a lasting investment in all forms of 'local capital'.

In addition, the SAFE-World database at the Centre for Environment and Society at the University of Essex tracks case studies in organic agriculture, providing a valuable reference to projects in developing countries. There has also been growing attention to the common ground between rural development and organic agriculture, chiefly around the issues of food security and environmental sustainability.

FAO's overview papers noted the connections between organic agriculture and biodiversity and between organic agriculture and ecoforestry in protected area management, describing the links between sustainable natural resource management and the food supply. While more systematic analysis is necessary, it appears that organic agriculture can make a unique contribution to improving access to food within market systems under certain conditions, and more significantly outside mainstream markets where most hungry people are found.

ACCESSIBILITY COMPONENTS

Productive Resources

The essence of organic agriculture is its reliance on local assets. These include productive resources such as land, water, forests, biodiversity, and environmental services such as pollination, predation, and nutrient cycling. Access to land includes tenure or the ability to secure long-term lease or ownership of productive growing areas to improve soil fertility and other conditions through organic methods.

In turn, organic methods can increase efficiency and returns from the land, thus assuring farmers' ability to maintain tenure over time. This is the case made by the Southeast African-American Organic Farmers network, a new farmer organisation in the southeastern United States.

Access to water implies sufficient clean water for agriculture and for food preparation and processing. Agroforestry provides collecting areas for food and food ingredients, and therefore plays an important role in facilitating access to food. It also serves two ecological roles: providing habitat for beneficial insects and birds; and enhancing water retention as degraded land is recovered. Biodiversity includes indigenous wild animals, plants, birds, insects, and fish; it also incorporates agricultural genetic resources and managed aquaculture stocks.

Environmental services are the ways that organic agriculture is able to benefit from existing natural processes while conserving these processes for future food production.

One example is the control of pests with beneficial insects instead of relying on costly and toxic pesticides, which kill beneficial insects as well as targeted pests. By investing in soil fertility, utilising local resources, and working with rather than against natural processes, organic agriculture increases the cost-effectiveness and resilience of food production. Water efficiency and crop diversification are also essential in this system. The result is reduced risk, which is more critical to food security than are maximum yields.

Household and community impact

When farmers have access to a broad range of productive resources and use organic systems, they are able to increase total yields, diversify their crops and improve their household incomes while ensuring the long-term sustainability of essential resources. Greater productivity coupled with lower input costs and reduced debt mean higher net incomes for households. Family and community members can meet the greater need for labour, increasing total community income.

One improvement supports another, drawing on farmers' creativity, traditional practices, and leadership to direct research and other supporting resources towards the most urgent local needs. A growing body of case studies supports the link between the introduction of organic practices and an increase in total farm yields, with increased access to food. Among them is the Tigray Project in Ethiopia, initiated by the Institute for Sustainable Development (ISD). Based on the principles of farmer control and ecological land management, ISD has created a local food system that uses diverse crops, produces high yields, optimises livestock production, utilises indigenous biodiversity and draws upon both indigenous and scientific knowledge.

As a result, total farm yields, diet quality, food access, and net incomes have all increased among thousands of farmers in more than 80 villages. The project also illustrates two less-often noted elements of long-term change: the institutionalisation of new practices through the participation of stakeholders and the importance of advocacy for farmers' rights, such as that to protect biodiversity. Another example of diversified, integrated, and intensified organic agriculture is the World Fish Centre fish pond and home crop production system in Malawi and elsewhere.

Access to food—including fish, with its unique health benefits -has been increased dramatically. The project features farmer control over fish genetic-stocks, on-site evaluation of resources and the continual on-farm mapping of resource flows and collaborative efforts of farmers and scientists. Another good example of organic agriculture strengthening local food security and access is the Thailand-based Green Net rural development programme which uses organic agriculture as an organising platform.

Family members must make decisions together regarding conversion of farms to organic production, and they must commit to enhancing and maintaining food plots as a requirement for organic certification. Family members are also trained, and in turn train others throughout the community to enhance the project's impacts.

Positive impacts

Farmers with access to productive resources can produce positive impacts far beyond their household and communities, including the stewardship of both wild and agricultural biodiversity. Biodiversity for food and agriculture encompasses all wild plants and animals, including trees, birds, fish, insects, fungi, soil microorganisms, and other organisms in or near agro-ecosystems as well as domesticated crops and animal breeds that have been selected for use in agriculture.

Genetic resources are an irreplaceable foundation for adaptation and improvement in agriculture. Use of genetic diversity is a time-tested strategy for reducing the risk of crop failure, as some varieties and crops are more resilient than others. Different crop varieties may also serve different dietary functions.

While conventional agriculture tends to focus only on the highestproducing varieties of plants and animals, considerably reducing the genetic base of agro-biodiversity worldwide, organic agriculture's emphasis on adaptation places high value on locally-adapted stocks because of their ability to succeed in specific micro- and macro-environments.

Organic practices produce a profound external benefit: the conservation of diversity. It is only in organic agriculture in which identification, use, evaluation, and conservation of a range of locally adapted, hardy, heirloom varieties of plants and animals is encouraged. Diversity must be used in order to be saved: only through it use can its value be demonstrated and it can continue evolving to fit new agricultural needs and environmental changes. This is especially significant at a time when an estimated 20 percent of the world's domestic animal breeds are at risk of extinction.

The important link between farmers and genetic conservation is recognised by the International Treaty on Planet Genetic Resources for Food and Agriculture. Ratified in 2001, the Treaty recognises the contributions made by farm communities in conserving and improving plant genetic resources. It also recognises their rights to participate equitably in benefit-sharing and in national decision-making about plant genetic resources. MASIPAG, a farmer-scientist partnership based in the Philippines, makes genetic conservation a high priority within its sustainable development programme. Through its collection, identification, multiplication, maintenance, and evaluation (CIMME) programme, indigenous varieties of rice, maise, and some vegetables are conserved directly by farmers.

Livestock conservation is also underway, including the improvement of locally adapted stocks. The organisation's description includes this statement: "With the practice of organic agriculture, the full genetic diversity of rice can be conserved, along with the related indigenous knowledge and cultural practices. This diversity is the heritage of farmers' communities, who can continue the work handed down by generations; meeting their local nutritional, social, and economic needs through breeding and selection". One benefit of using indigenous genetic resources is their potential for superior productivity under low-input conditions.

Selection of locally adapted seeds and breeds for increased production is more effective than selecting imported, productive varieties for adaptation. Traits for adaptation respond relatively slowly to selection, requiring many generations to become fixed; this is especially true of adaptation to extreme environments. An example is the Gulf Coast native sheep of the southeastern United States. This breed, derived from imports of Spanish sheep beginning in the 1500s, has been shaped by natural selection to tolerate an environment that is usually very difficult.

These sheep demonstrate heat tolerance, parasite and foot-root resistance, reproductive efficiency and longevity under low-input management, including low-quality forage. Selection is now underway to improve meat performance while retaining environmental adaptation. Preliminary experience with Damara sheep, indigenous goats, and Nguni cattle in South Africa has led to similar conclusions. A contrasting example comes from Cuba, where had a dairy herd composed of North American Holsteins, which had been selected for high production under high inputs was raised in the 1980s.

When a shift to low-input, organic farming had to be made in the1990s, these cattle could not survive, much less be productive. In Indonesia, indigenous rice genetic diversity was nearly lost when planting of local varieties was forbidden in 1965. The NGO Pusspaindo has been working with farmers since the 1990s to locate, conserve, and evaluate local varieties of paddy rice, as well as to encourage cultivation using organic methods. Observation suggests that locally adapted varieties can be very productive. The organisation also collects and shares indigenous knowledge on the production and processing of this rice, and on the cultural traditions surrounding it.

Knowledge Systems

It is critical to understand that organic agriculture is a knowledge-based approach to agricultural problem solving as opposed to the product- or input-based approach of conventional systems. Organic agriculture requires asking specific questions such as 'how has this particular pest or disease become a problem?', then searching for the answers. It requires an understanding of farms as entire systems and farmers as capable experimenters and innovators with a wealth of knowledge and experience.

Organic agriculture should not be seen as a recent creation from northern countries, but as the ongoing adaptation of traditional knowledge and its validation internationally with modern science. The agro-ecological science to draw out critical patterns in this process is still in its infancy. Jules Pretty and colleagues use the term 'human capital' to describe the "total capability residing in individuals, based on their stock of knowledge, skills, health, nutrition ...Peoples' productivity is increased by their capacity to interact with the productive technologies and with other people. Leadership and organisational skills are particularly important in making other resources more valuable".

In the conventional agricultural model, knowledge is seen as the domain of off-farm experts instead of being held by farmers. It is tightly focused around: the most productive plants and animals; a specialised set of inputs; and the free flow of fuel and water. Emphasis is placed on the system's transferability and local genetics are generally dismissed.

Farmers in conventional systems are often labelled as 'end users' of the technology transferred, and existing knowledge is seen either as irrelevant or an obstacle to progress. As a result, knowledge in the conventional system tends to be hierarchical, deep but narrow and site-general rather than site-specific.

Increasing farmer and market dissatisfaction with the rising environmental costs of conventional agriculture and economic stresses are helping to drive the growth of organic markets, leading to the conversion of millions of hectares to certified organic farmland, permanent pasture and collecting areas.

The continued expansion of organic agriculture is still limited in most countries, however. This transition has been made primarily by individuals and communities, with little or no institutional support from governments or the agriculture establishment. NGOs have been instrumental in training farmers as part of organic agriculture outreach programmes or, increasingly, as part of poverty-reduction efforts.

The chief obstacles to expansion continue to be lack of knowledge about organic methods such as intensive cultivation, lack of information about certification and market access, and lack of site-specific research to address farmers' needs. When NGOs from the North first brought information about certified organic agriculture to the South, it was based on formal certification standards and practices designed for their own temperate regions. This posed many challenges to developing country farmers.

While organic systems have always emphasised the principle of reliance on local resources, organic agriculture must be further adapted to the resources of the tropics and sub-tropics in the developing world. This requires the continual involvement of farmers, local NGOs and universities, which have their own baselines of knowledge and experience. Partnerships of these groups, in the form of peer-based and farmercentered programmes, are allowing organic agriculture to live up to its mission as a locally based system and evolve in all parts of the world.

Organic agriculture continues to evolve into a broad, diverse, and geographically adapted system. The concepts of participatory and grower-group certification systems are prime examples of continuing innovation, which are helping to mitigate some of the financial and market barriers faced by small farmers in developing countries. Such approaches can overcome economic barriers by applying local skills, knowledge and organisational development to facilitate access to much-needed local and valuable northern market opportunities. However, attaining international acceptance of these innovative strategies remains a challenge.

The Centre for Indian Knowledge Systems, an India-based NGO, describes this goal with the following statement: "We strongly believe that the future lies in understanding and harnessing the potential of indigenous knowledge systems and integrating them into the mainstream of scientific, industrial and everyday thinking. Only then can we build a strong and self-reliant society".

Household impact

One way to access to food is for more people to become farmers and gardeners, growing more food where it is needed. This is being accomplished by the creation of new intensive agriculture projects through investment in small-scale gardening in urban and rural areas. Training in organic practices—such as increasing soil fertility, diversifying crops, intensifying production, and incorporating poultry, rabbits and other micro-livestock, goats, and even dairy cattle—is particularly beneficial to resource-poor households, especially women who only have access to kitchen gardens.

Total yields are increased and may even include surpluses, which can be sold to generate income. In addition, when women are given opportunities to acquire the skills and knowledge in food production, they are able to assume more control over the household food supply and contribute their own skills and creativity to increasing it.

In Chile, organic vegetable garden systems developed by the Centre for Education and Technology (CET) and by CET SUR, have demonstrated that even small gardens can be used to achieve year-round household food self-sufficiency, surpassing the productivity of conventional farms. In addition, heirloom seeds of the Mapuche people are being used, conserved and shared, along with supporting indigenous knowledge. In Havana, Cuba, urban organic gardening projects led by the Grupo de Agricultura have increased access to food through the intensive use of garden spaces.

During the financial collapse of the early 1990s, thousands of gardens were constructed in Havana. Technical support and gardening supplies, including seeds, plants, fruit trees, and compost, were provided by the Cuban Government. These gardens were organic by necessity as there were no chemical inputs available. Kitchen, neighbourhood and community gardens produced approximately one half of the country's fresh vegetables, 38 percent of its non-citrus tropical fruits and much-needed employment for thousands of Cubans.

Community impact

While increasing families' access to food remains critical, often the combination of farmer-to farmer exchange and community-wide organising catalyses impact. Community-wide strategies to increase access to food are highly effective approaches that can build on local strengths, problemsolving skills, and experiences of participating farmers and their families to accelerate success. Actions may be formal though organised groups or informal through common activities focused on a local demonstration centre; both have transformative effects.

As IFAD's work in Asia has shown, it is the existence of farmer groups that makes or breaks the success of organic projects. These important organisations are platforms for sharing site-specific knowledge and experience, demonstrating and training in new techniques, and engaging in locally based research. The crucial role of communities is best illustrated by the efforts of potato farmers in the mountains of Northern Potosi, Bolivia to improve their livelihoods and maternal and child health.

Working in partnership with the United States-based NGO World Neighbours, through its Bolivia Integrated Development Program, farmers were able to initiate research on indigenous potato varieties, including seed-size variations, spacing and cultivation techniques. They also adapted findings from another World Neighbours project in Honduras on the use of mucuna (velvetbean) to increase maize yields. While velvetbean does not grow in high altitudes, the farmers identified lupin as an alternative.

Potato yields increased dramatically with the use of lupin, rising still further when sheep manure was added.

Yields for the 2 000 participating farmers in 198 communities increased to the extent that cultivation areas could be reduced. In Pakistan, the Sindh Rural Women's Uplift Group has an Organic Fruit Production programme based at the Panwhar Farm.

The Group owns a 43-hectare fruit orchard that is managed organically and produces high yields of citrus and mango. According to Farzana Panhwar, Managing Director of the group, the high quality of produce has enhanced its price, and more income has resulted in a two to threefold increase in community food security. Innovations in fruit varieties and post-harvest handling are expected to further increase the value of the farm's crops. Training and employment of women have also increased.

Regional and National Impact

In many regions, there are a few tenacious problems that must be resolved in order to improve food security over a large area. Control of pests is a particularly important issue for organic agriculture since local resources are expected to replace the use of broad-spectrum pesticides with the same or improved effectiveness at lower cost. Research conducted in Kenya and Uganda by African Insect Science for Food and Health (ICIPE) has achieved striking results with the vutu-sukumu ('push-pull') pest management system. The system was driven by farmers' needs, the research utilised a multi-disciplinary approach and its application was enlightened by traditional practices.

Stem borers and striga are pests affecting maize, millets, rice and sorghum. These pests may account for maize losses of 15-40 percent in eastern Africa and 10-20 percent in southern Africa. When they occur together, they can cause the loss of an entire crop, threatening the food supply of 27 million people in the region. With the 'push-pull' approach, cereal grains are intercropped with silver leaf and molasses grass, which repeal or 'push' the stem borers.

Trap crops such as napier grass and Sudan grass, which are even more attractive to the stem borer than the maize or other crops, then attract and 'pull' the pests. This strategy creates significant increases in yields by reducing losses. Desmodium, a nitrogen-fixing plant suppresses striga and increases soil fertility. In addition, these plants are useful as fodder, which improves livestock health and enhances child health through increased milk production. ICIPE has trained a network of farmer-teachers, and over 3,000 farmers have begun to replace expensive pesticides with low-cost, knowledge-based, culturally appropriate practices based on an understanding of natural systems.

Access to markets

Access to markets is essential for all farmers, as surplus crops can be sold to generate income. Domestic community, peri-urban and regional organic markets are expanding in countries such as Brazil, China, Egypt and India. In China, there is a rapidly growing market for 'green food', which is not organic, but grown with reduced chemical inputs.

As Richard Sanders writes, "Given that these forms of agriculture reduce farmers' use of chemicals compared to conventional farming—chemicals which are costly to produce and environmentally degrading to use—they contribute to ensuring a more environmentally sustainable future for Chinese farming, post-WTO entry, whilst providing opportunities for farmers to enrich themselves at the same time: a 'win-win' state of affairs. This will remain the case, however, only so long as the state is prepared to create and reinforce appropriate institutional arrangements".

In most developing countries, however, the international markets of Europe, North America and Japan have attracted the most interest. Gaining access to these markets—and the promise of fair prices and stable, long-term incomes—has made trade essential to improving incomes and livelihoods of many organic farmers. The challenge is assuring that this approach is balanced by the strengthening of local production and consumption cycles. Access to global markets is a multi-step process requiring information and experience about crop and product selection, certification for import markets, quality standards, market timing, trade regulations, marketing contracts and several other factors.

For this reason, success is best assured when farmers form groups that can advocate for their interests. This can also be accomplished through partnerships with NGOs that have legitimate trade experience, and through fair contracts with trading companies, food processors and other buyers. Both domestic and global markets are important and can play complementary roles. As demonstrated by East Africa-based EPOPA and Egypt-based SEKEM initiatives, it is possible to use organic agriculture to increase and diversify local food production, support local food supplies and tap market opportunities for value-added products. The key is to strike a balance these two markets.

As markets support organic expansion, there is often a ripple effect, spurring development of related enterprises such as those that supply compost, bio-pest and disease control stocks, seeds, plants, and other inputs, as well as crop-handling and processing facilities. These may not always be available, and they may be economically efficient only as an input-substitution transition before a resilient ecosystem is established that allows farmers to count on ecosystem self-regulation. Such 'green' businesses share the benefits of both domestic and international organic markets.

Currently, Europe and the United States account for over 80 percent of the rapidly growing organic market, now valued at US$40 billion. This represents a significant market opportunity for farmers in developing countries as demand for organic tropical foods, medicines and other products continue to grow. However, many challenges still exist, including the need to build physical and information infrastructures, high travel costs, long distances to markets and the lack of adequate financial supports.

In addition, since organic prices are neither institutionalised nor standardised, small farmers remain vulnerable to conventional downward pressures of market supply and demand, and the affects of market consolidation. Stimulating local market demand represents one of the most important long-term strategies for strengthening local food security and ensuring more equitable access to food. This can also reduce risk and support more stable and less volatile marketing of products. Organic agriculture also enhances development of enterprises such as agro-tourism.

For example, the success of the organic spice trade in Kerala, India has attracted the support of the State Government, which is now promoting organic farming through a programme called Organic Sustainability of Kerala. The tourism industry has in turn provided support, realising the opportunity of branding Kerala as a 'safe food destination'.

There are several organic agriculture and eco-tourism projects in Central America, including: El Triunfo Biosphere Reserve in Mexico where organic coffee production complements biodiversity conservation; organic cacao agroforestry in the Talamanca-Caribbean Biological Corridor in Costa Rica where the transition to certified organic cacao has raised incomes, reduced pressure on the environment and slowed habitat loss; El Impossible and Los Volcanes National Parks in El Salvador where organic and shade coffees are being produced; and the possibility of organic agriculture development near the Monte Verde Preserve in Costa Rica.

It takes capital to develop new agriculture infrastructure such as roads and credit and savings systems, and enterprises such as processing and marketing systems. Much work remains in these areas, especially for 'brick-and-mortar' supports and training and education for farmers, bankers, lenders and rural development workers. Access to fair credit and support for locally owned regional food supplies and processing are essential to ensuring long-term local viability, economic stability and food security.

While this true of all agricultural infrastructures, in organic agriculture these improvements benefit a broader community, with positive impacts on food and economic security. While markets are most often thought of as opportunities for export, it is also necessary to consider the affects of imports of food on local production. Current trading practices may be based on predatory export/import policies such as the importation of subsidised foods from Northern markets, which undermines the ability of local farmers to obtain fair prices for their own crops.

As a result, local production and market infrastructure is damaged along with local capacities to attain long-term food security and self-sufficiency. The potential of organic agriculture to produce more diverse crops with greater economic sustainability may protect countries from becoming trapped in such policies -if there is the political will to protect farmers and communities.

Household impact

Maintaining family-wide, fair access both to the means of food production and to food for consumption is a critical component of food security. Organic projects that increase household income from new markets—while maintaining local household food production—represent a win-win trade scenario. To be fair, trade must be based on mutual positions of strength and need; it must raise incomes and enhance food security at both ends of the trading relationship. With access to fair trade, household income can increase along with higher prices for indigenous crops that are inexpensive to grow or that grow only in a few regions.

In Bolivia, Peru, and other Andean countries for example, farmers are growing organic quinoa for export. Quinoa is a highly nutritious, gluten-free grain used in a variety of processed foods in place of wheat or rice for the very rapidly growing gluten-free market. The high-altitudes of the Andes are ideal for quinoa, which does not thrive in other settings. In Kenya, the Mount Kenya Organic Farm has improved food production for household and community supplies while producing certified organic borage for export. This crop is inexpensive to grow, but its seeds can be cold-pressed into highly valued neutraceutical oil for export to Europe and the United States.

Farmer groups market the product to Earthoil Kenya Ltd, generating an estimated US$64 000 during the pilot year for the Nanyuki community of 80 smallholders. It is hoped that a market for organic honey will also be developed, as borage attracts large numbers of bees. The growth of the global market for organic medicinal herbs, many of which are indigenous to developing countries, creates another attractive opportunity for smallholders...

Community impact

The most effective strategy for creating and optimising market access is for small farmers to organise into cooperatives or other democratically run grower groups. These groups can leverage community production into a single unit and then advocate from a position of greater strength for appropriate certification, favourable contracts and fair prices. The organic grower group model allows for highly organised groups of farmers to save on costs by establishing internal organic certification systems, in which they pay only one certification fee for the entire group rather than for each of the individual members.

Over 100 000 limited-resource organic farmers in Africa, Asia and Latin America have accessed northern markets through this creative strategy, many for the first time. This has translated into over US$500 million in organic retail sales in the United States alone for such products as organic coffee, chocolate, and sugar. Partnerships between farmer organisations and NGOs are often especially powerful in creating mutually beneficial trade relationships and assuring that internal-control systems are fully functional.

Winning global acceptance and organic accreditation is the key to expansion of this effective but challenging approach. The Uganda-based Export Promotion of Organic Products from Africa (EPOPA) initiative has been a model in developing global markets for farmers in East Africa through the private sector. Founded in 1994 by the Swedish International Development Agency (SIDA) and implemented by Agro Eco and Grolink, consultancies based in the Netherlands and Sweden, EPOPA's mission is 'Development Through Trade'.

Within the initiative, partnerships are created to link small farmers with global market opportunities. EPOPA provides the knowledge of trade processes and market potential, training farmers in crop production and handling; the farmers supply knowledge of local crops and techniques. Premium prices obtained for products enhance farmers' incomes. Organic practices used for export production are also improving household and community food production through higher yields and better quality food crops with resilience to environmental and financial risk.

Regional and national impact

Facing the challenges of certification and market access, farmer and other partnering organisations have pioneered participatory certification systems which have significantly impacted organic producers around the world. The challenges include growing costs, complexity and burden of organic certification schemes that are tailored to international markets. Community members are trained in certification techniques, reducing external and third-party certification costs so that resource-poor farmers can access organic markets they could not otherwise afford.

New legislation in Bolivia, Brazil, Costa Rica, and India is allowing accreditation of participatory guarantee systems for the domestic markets in these countries. This has significant potential for farmers in Brazil and India, where the domestic markets are large and increasing rapidly. This model is also an effective way of organising community support systems to re-link farmers with consumers. For example, the Brazil-based Ecovida describes its innovation as a participatory guarantee system for ecological and organic foods.

The network is characterised by group problem solving and communication, building upon the local certification system developed by French winemakers in early 1900s, who wanted to distinguish their products from those of less superior quality. Ecovida's principles include commitments to social equity, solidarity, cultural values and community autonomy. Its goal is to create an environmentally and financially sustainable system that supports both local and global markets. One of its major challenges is to achieve recognition and accreditation by Northern governments, certifiers and their markets.

REFERENCES

Fowler, C. & M, P., *Shattering: Food, Politics, and the Loss of Genetic Diversity*, University of Arizona Press, Tuscon, 1990.

Parrott N. and T. Marsden, *The Real Green Revolution – Organic and agroecological farming in the South*, Greenpeace, 2002.

Sanders, R., "A Market Road to Sustainable Agriculture? Ecological Agriculture, Green Food, and Organic Agriculture in China," *Development and Change* 37(1), 2006.

Scialabba, N. & Hattam, H., "Organic agriculture, Environment, and Food Security," *Environment and Natural Resources Series No. 4*, FAO, Rome, 2002.

Von der Weid, J., "Agroecology as the Way to Achieve Food Security," *FAO International Conference on Organic Agriculture and Food Security*, May 3-5, 2007, Rome.

4

Methods of Organic Crop Production

The USDA defines Organic Agriculture as "a Production system that is managed to respond to site-specific conditions by integrating cultural, biological, and mechanical practices that foster cycling of resources, promote ecological balance, and conserve biodiversity."

More specifically, organic farming entails:

— Use of cover crops, green manures, animal manures and crop rotations to fertilise the soil, maximise biological activity and maintain longterm soil health.

— Use of biological control, crop rotations and other techniques to manage weeds, insects and diseases.

— An emphasis on biodiversity of the agricultural system and the surrounding environment.

— Using rotational grasing and mixed forage pastures for livestock operations and alternative health care for animal wellbeing.

— Reduction of external and off-farm inputs and elimination of synthetic pesticides and fertilisers and other materials, such as hormones and antibiotics.

— A focus on renewable resources, soil and water conservation, and management practices that restore, maintain and enhance ecological balance. Many organic farmers, including Wende Elliott and Joe Rude of Colo, Iowa, view organic production as a means to work with the environment and maintain the balance of their ecosystem. "Natural systems work hard if you incorporate biodiversity into your operation instead of fighting it," said Rude, who co-farms 125 acres of pastured poultry, corn, hay and alfalfa.

Using nature as a model for the agricultural system—recycling nutrients, encouraging natural predators to manage pests, increasing plant densities to block weeds—organic farmers don't merely substitute nontoxic materials for pesticides and fertilisers, but rather consider the farm as an integrated entity, with all parts interconnected.

When livestock and poultry are incorporated into organic systems, the potential for diversification and integration is even greater: Livestock feed on grasses and mixed forages, both of which help improve soil structure. At the same time, livestock provide manure to fertilise soil, and can be used to "cull" any nonharvestable crops.

Elliott and Rude, like many organic farmers, want to raise food free of hormones, antibiotics and pesticides. For many years, organic producers and proponents have claimed that organic farming is gentler on the environment. Research now confirms this:

— The Sustainable Agriculture Farming Systems (SAFS) project at the University of California-Davis, a 12-year research station experiment comparing conventional and organic systems, showed water infiltration rates to be 50 percent higher in the organic system. The project, supported by a grant from USDA's Sustainable Agriculture Research and Education (SARE) program, also showed that the organic system had one-third the amount of water movement into surface and groundwater as the conventional system. The organic system was more efficient at storing nitrogen and had positive effects on soil quality, including higher biological activity and a doubling of organic matter in 10 years.

— An organic cropping system consumed three to four times less energy than a conventional system, while also producing six times more biomass per unit of energy consumed in a South Dakota State University comparative trial at the Northeast Research Station near Watertown.

— A SARE-funded study evaluating pesticide and nutrient loads in subsurface drainage on organic and conventional farms in Illinois found less nitrate, chloride and atrasine in the water draining from the organic fields. More recent research also shows that organic farming systems can be equally productive and economically competitive with conventional systems, and in some cases, more resilient. Consider that:

— A study comparing longterm established organic and conventional tomato farms in California's Central Valley found comparable yields.

— An article published in the Organic Farming Research Foundation Bulletin reviewing data from seven universities and two research station experiments verified that organic corn, soybean and wheat yielded, on average, 95 percent of conventional.

— Many studies have shown that organic systems perform better than conventional ones under drought conditions.

ORGANIC FARMING IN THE U.S

J.I Rodale, founder of the Rodale Research Institute and Organic Farming and Gardening magazine, is commonly regarded as the father of the modern organic farming movement. Beginning in the 1940s, Rodale provided the main source of information about "nonchemical" farming methods and was heavily influential in the development of organic production methods.

Rodale drew many of his ideas from Sir Albert Howard, a British scientist who spent years observing traditional systems in India. Howard advocated agricultural systems reliant upon returning crop residues, green manures and wastes to soil, and promoted the idea of working with nature by using deep-rooted crops to draw nutrients from the soil. By the 1970s, increased environmental awareness and consumer demand fuelled the growth of the organic industry. However, the new organic industry suffered growing pains.

Although there was general agreement on philosophical approaches, no standards or regulations existed defining organic agriculture. The first certification programs were decentralised, meaning that each state or certifying agent could determine standards based on production practices and constraints in their region. An apple farmer in New York has very different challenges than an apple farmer in California, for example. The downside of this decentralised approach was a lack of clarity about what "organic" meant from state to state.

A movement grew to develop a national organic standard to help facilitate interstate marketing. In response, Congress passed the Organic Foods Production Act (OFPA) in 1990 to develop a national standard for organic food and fiber production. OFPA mandated that USDA develop and write regulations to explain the law to producers, handlers and certifiers. OFPA also called for an advisory National Organic Standards Board to make recommendations regarding the substances that could be used in organic production and handling, and to help USDA write the regulations. After years of work, final rules were written and implemented in fall 2002.

Although the actual production techniques of organic food have not changed dramatically since the implementation of the national standards, "organic" now is a labelling term that indicates that food has been grown following the federal guidelines of the Organic Foods Production Act.

The national standards also specify that any producers who sell over $5,000 annually in agricultural products and want to label their product "organic" must be certified by a USDA-accredited agency. Companies that process organic food must be certified, too. Any farms or handling operations with less than $5,000 a year in organic agricultural products are exempt from certification. Those producers may label their products organic if they follow the standards, but they are prohibited from displaying the USDA Organic Seal.

ORGANIC CROP PRODUCTION

Desiging a farming system to tie togather principles of sustainability and productivity is complex. Organic farmers must consider how the various components of their system—rotations, pest and weed management, and soil health—will maintain both productivity and profitability. Although practices vary from farm to farm and region to region, at the core of any successful annual organic farming system is the crop rotation. According to "Cereal-Legume Cropping Systems: Nine Farm Case Studies in the Dryland Northern Plains, Canadian Prairies, and Intermountain Northwest," productive rotations:

— Enhance soil conservation and build soil organic matter;

— Provide weed, disease and insect control;

— Enhance water quality and conservation, biological diversity and wildlife habitat; and

— Ensure economic profitability for the farming system.

As the main management tool for all aspects of the farming system—including weeds, pests, insects, soils, and crop production—a well-planned rotation is more than the sum of its parts, addressing the connections between all of those factors. For example, successful rotations, according to "Switching to a Sustainable System" by Fred Kirschenmann:

- — Include the use of cover crops to provide fertility, control weeds and provide habitat for beneficial insects;
- — Have a diversity of plant species to encourage natural predators, discourage pest and disease build-up, and minimise economic and environmental risk;
- — Provide a balance between soil conservation and crop production by adding organic matter to the soil to both supply nutrients and improve soil quality properties such as water infiltration and water holding capacity; and
- — Provide weed control by alternating between warm and cool weather plants and including weed inhibiting plants.

Newark, N.Y., organic vegetable farmer Elizabeth Henderson, who farms 15 acres, considers rotations featuring summer and winter cover crops a key component of her successful system and relies on them to minimise erosion, maintain and build soil quality, and control pests. For agronomic crops, a standard organic corn belt rotation of alfalfa, corn, soybeans and small grain accomplishes multiple functions because:

- — The legumes fix nitrogen, providing for the subsequent non-legumes in the rotation;
- — Several pest cycles are interrupted, especially that of the northern and western rootworm species, which can be devastating to corn;
- — Several plant diseases are suppressed, including soybean cyst nematode; and
- — Weed control is enhanced when perennial weeds are destroyed through cultivation of annual grains; most annual weeds are smothered or eliminated by mowing when alfalfa is in production.

For some farmers, switching to an organic rotation may not be more difficult than expanding upon or changing the timing in an existing rotation. When Lydia and Dennis Poulsen of Snowville, Utah, decided to convert their 800-acre beef, hay and small grain operation to organic, making the switch was much easier than expected.

"An organic dairy was coming and they needed feed," recalled Poulsen. We had alfalfa in our original rotation and we were already trying alternatives to make the ground healthier. Because their alfalfa-wheat-oat hay rotation fit right into an organic system plan, the only substantial change they made on their ranch was to plow in alfalfa as green manure for their subsequent wheat crop, rather than letting the cows mow down the alfalfa in its final year.

Soils

Along with developing a suceesful rotation, ensuring healthy soil is imperative to a profitable and successful organic system. "A lot of people don't think of the soil as an ecosystem but, in fact, it's probably the most complex ecosystem on earth," said Ray Weil, a University of Maryland soil science professor. "A healthy soil should be breathing out carbon dioxide, breathing in oxygen. It should hold and absorb water so the plants can survive between rains. It should resist erosion." By contrast, a less healthy soil can wash away and pollute surface waters. From a production standpoint, poor quality soil can limit plant growth and vigor.

In organic farming systems, the majority of nutrients are supplied from organic matter additions such as compost, manures and cover crops These amendments not only feed the plants, but the soil organisms as well. As soil organic matter accumulates, soil structure improves, and populations of other important soil organisms, such as earthworms—which tunnel through the soil, improving aeration and infiltration increase. Those organisms break down organic material to release nutrients at a steady pace so they are available for plant uptake.

Soil microorganisms also hold nutrients in a more stable form so they are less susceptible to being lost—through leaching, soil erosion or runoff. The soil is a virtual microscopic zoo of organisms. Soil biologists are just beginning to tease apart how those organisms function in organic farming systems. Numerous studies show that organic systems have higher microbial populations and activity.

The longterm SAFS trial in California's Central Valley comparing organic and conventional farming systems in a tomato, bean, corn and safflower rotation found significantly higher microbial populations and activity in organic systems than the conventional ones. New research from North Carolina State University shows that increases in microbial populations and microbial activity may occur by the first or second year of the transition to an organic system. Researchers also are discovering that they can improve fertility in organic systems by micro-managing the soil fauna.

In the SAFS experiment, researchers studied the role of bacteria-feeding "good" nematodes, small soil organisms that help make nitrogen available to plants. The researchers found that by irrigating plots in the fall to improve cover crop germination, the nematode population increased. This higher beneficial nematode population led to more nitrogen release from the cover crop in the spring. The nematodes also stored nitrogen over the winter that might otherwise have been lost. Cover crops, an essential part of organic systems for soil building and soil fertility, also benefit the soil by improving soil structure, which in turn improves water infiltration and water-holding capacity.

The longterm systems trial at UC-Davis proved some of those benefits dramatically, such as 50 percent higher water infiltration and 35 percent lower runoff in the organic plots. "Nobody could have possibly predicted such a dramatic difference in the water runoff and infiltration between the organic and conventional systems," said SAFS project leader Steve Temple. "It's given us a new appreciation of the importance of cover cropping and residue management."

Cover crops planted after a crop is harvested—also known as catch crops—recover nutrients that would otherwise leach into the subsoil and groundwater. Cover crops prove invaluable to organic growers who don't have access to affordable sources of compost and manure. A study of potato production in Idaho found that legumes such as alfalfa, pea and pea-oat hay could provide 80 to 100 percent of nitrogen needed for a potato crop, and if harvested for feed or seed, 40 to 60 percent of the required nitrogen for the subsequent crop.

Similarly, a northern California research project showed a nitrogen replacement value of 150 pounds per acre with cover crops. Carmen Fernholz, who grows organic barley, oats, wheat, flax, corn, soybeans and alfalfa on his 410-acre west central Minnesota farm, manages a three-year, four year or even longer rotation heavily reliant upon cover crops. Without exception, he underseeds all of his small grains with a legume crop, such as red clover or annual or perennial alfalfa.

After harvesting the small grain, he allows the underseeded legume cover crop to serve as a green manure—or, with perennial alfalfa, as a cash crop. The number of seasons for the perennial alfalfa will depend on the weed and nutrient histories of the particular field.

Organic farmers also use manures and composts regularly, especially when they are accessible and affordable. Many organic farmers make their own compost, either by using livestock manure from their own operations or from a nearby source and combining it with straw or wood shavings. Manures and composts provide many of the same soil-building benefits as cover crops.

Weed Management

Although they joke about it now, it probabiy didn't seem too funny when Joe Rude and Wende Elliott thought they'd lost their first year alfalfa crop to an invasion of weeds. Yet weed management for organic and transitional farmers is a formidable problem and ranked number one among research priorities in a national survey of organic farmers. Standard organic weed management strategies include:

— Smothering weeds with cover and forage crops. A dense mat of cover crops will prevent weed germination or crowd out weeds struggling to gain a toehold. Residue from a grass cover crop decomposes slowly, while legume residues break down faster. Grass-legume mixtures also can control weeds while providing more nitrogen to the cash crop. Forages serve a similar process; in his five-year alfalfa, corn and oat rotation, "the three year alfalfa goes a long way to getting the weed seed bank out of the field," said Joe Rude.

— Managing weeds selectively. Identify weeds and manage them according to their lifecycle and reproductive strategies. For example, tilling weeds like quack grass and Canada thistle is ineffective in the short run, since tillage may propagate their rhizomes. Repeated cultivation, however, forces them to draw upon their storage, and can eventually weaken the population. Biennials, on the other hand, must not be allowed to seed and persistent mowing can eventually exhaust root reserves.

— Conservation tillage. Mark Davis, an agronomist working on a longterm organic farming systems research trial run by the USDA Agricultural Research Services Sustainable Agricultural Systems Lab in Beltsville, Md., uses a regular no-till planter to sow corn directly into a living stand of vetch, and then rolls and crimps the vetch with a Buffalo Stalk chopper. "This system has great potential for no-till organic cropping systems because it provides weed control and nitrogen fertility at the same time," Davis said. By allowing the vetch to grow longer in the spring, the researchers are increasing the amount of nitrogen added to the system. At the same time, the mat of vetch smothers the weeds long enough for good early season weed control. "It gives you the best of both worlds," Davis said. "Your no-till practices help reduce soil erosion, improve soil structure and increase organic matter, and you can still manage the system organically." The one caveat, he cautions, is that in years with a dry spring, the late growth of the vetch can deplete soil moisture for the following crop.

— Using living mulches. Inter-seeding one crop into another can be done on a large scale by sowing rye from aircraft over corn acreage, or from tractors or by hand. The second crop, which should germinate after the first, will compete for nutrients and moisture so this technique should only be used when crops are well established or have ample soil fertility and moisture. Dutch white clover, for example, is effective in corn or late season brassicas. Its high density keeps out weeds, it fixes its own nitrogen, and it is low growing so it doesn't compete with the crop for sunlight.

Many organic farmers also use some sort of mechanical weed control in combination with the above strategies. Some organic farmers believe that weeds do not need to be eradicated, just managed. Knowing when a weed is a threat and when it can be ignored, something often gained by experience, remains a common strategy. "Our farm becomes so much simpler all the time," said Dan Nagengast, who farms five acres of cut flowers and mixed vegetables in Lawrence, Kan., and has been growing organically for 15 years. We've learned from our mistakes—

it used to be if we were eight to 10 days away from harvesting lettuce, we would hoe the weeds. Now we know when the crop will make it, and we don't have to do all the extra things we thought we needed to.

In a bit of a radical departure from the conventional approach of "the only good weed is a dead weed," some organic farmers choose to integrate weeds into their cropping systems for the benefit of the whole farm. Steve Gilman, who farms 15 acres in Stillwater, N.Y., and grows fresh market vegetables such as lettuce, tomatoes and peppers in 4-foot-wide raised beds, decided there was no need to spend the "time and energy needed to keep the two-foot wide, permanent strips between the beds clean-cultivated."

As explained in his book "Organic Weed Management," Gilman was concerned about the susceptibility of the bare soil to erosion. So he began planting "biostrips." First, he eliminated perennial weeds such as quack grass and thistle with repeated cultivations before forming the beds. Then he sowed Dutch white clover between them, allowing a mix of perennial grasses, wild herbs and wildflowers to flourish.

Insects and Diseases

"Organic farming require more intensive management than conventional methods," said Joe Rude, the poultry farmer from Colo, Iowa, "because without access to a broad spectrum of pesticides and antibiotics, you have to understand the life cycle of the pests that are attacking your crops and animals. You have to understand the biological and chemical processes and work with the environment." For organic farmers, this means employing strategies such as crop rotations, enhancing biodiversity, determining threshold levels of pest populations, introducing natural enemies and using good sanitation practices.

Although certain sprays such as Bacillus thuringiensis (Bt) and rotenone are permitted, many organic farmers do not rely upon them exclusively. Rather, the key focus of organic pest control is prevention. In California, where dry summers make it less challenging to grow organic fruit, a SARE-funded on-farm study comparing organic and conventional apples found that synthetic pheromones, biological control agents and sanitation successfully controlled codling moth mating in most locations.

Disease control with sulfur and copper, when timed right, was as effective in scab control as the synthetic fungicides used in conventional systems. the In New York, where the wet humid summers pose enormous challenges for organic fruit production, Robert Pool, a Cornell University viticulture professor, found that organic grapes could be managed to be as pest-and disease-free as conventional ones. During a three-year, SARE-funded experiment, which compared organic and conventional grape production on three varieties, pheromone disruption and insect scouting allowed researchers to eliminate regular insecticide use.

"Going in, we thought the main problem would be diseases," said Pool. But instead, researchers applied, on average, less than one spray per year and found that the powdery mildew commonly observed on Concord grapes was far less destructive than predicted. The researchers also learned that pheromones successfully controlled grape berry moth and that scouting allowed them to control insects that emerged when the regular spraying was eliminated. They avoided an expected huge spike in grape leafhopper by releasing predatory wasps.

Other SARE-funded research has shown similar control:

— In a study of potatoes in Idaho, researchers were able to control Colorado potato beetle with mineral and biological compounds.

— In the Northeast, where sweet corn can be devastated by caterpillar pests, Bt and corn oil were used effectively to control corn earworm. Eight farmers from Vermont to Connecticut found that the oil controlled ear damage in 83 percent of their trial plots in 2000.

Many organic farmers have observed that, over time, pest populations seem to decrease. Results from a California on-farm organic tomato experiment showed the presence of 46 percent more predators and parasitoids and 43 percent more natural enemies on the organic farms, which could provide one possible explanation for reductions in pest populations.

A SAREfunded study in Washington testing mowing frequency in pear orchards found that mowing only once a month rather than more frequently as done on conventional farms, creates alluring habitats, attracting beneficial insects that control pests. "By reducing the frequency to once a month, we see a dramatic increase in natural enemies moving into the ground cover without a big increase in pests that feed on fruit," said David Horton, the ARS researcher testing mowing regimens.

Stone fruit grower Marilynn Lynn of Bridgeport, Wash., relies on living mulches to attract beneficial insects that prey on potential pests before they can harm her peaches, apricots and nectarines. "We mulch extensively," Lynn said during a satellite broadcast about organic production aired by Washington State University in spring 2003. Calling their orchard grass, yarrow and clover covers a "bed and breakfast" for beneficial insects, she added: "They give a nice diversity to the floor of our orchard, providing food and water in the spring when they wake up."

For soil borne-disease control in organic systems, many growers use composts, long known as effective plant pathogen suppressants. Rotations also are important for decreasing pathogen populations, as most pathogens are plant specific. In general, rotating the crop, planting resistant varieties, and adding organic matter have all been shown to reduce the incidence of soil-borne diseases: In the SAFS project in California, a four-year organic rotation had lower incidence of corky root and red root rot than a two-year conventional rotation; an on-farm tomato study in the Central Valley of California showed that organically managed soils may be suppressive to the organism that causes corky root; and in North Carolina, another SARE-funded study showed disease was significantly reduced by organic soil fertility amendments and on organic versus conventional farms.

Livestock Systems

Traditionaly, livestock have played an important role in integrated operations and fit well in organic farming systems. Livestock feed on forages and grasses, essential elements of organic rotations, and provide manure, an important organic fertiliser. Although semi-confined livestock systems are allowed under the federal organic rules, the animals must be given access to fresh air, sunlight and the outdoors.

Most organically raised animals do have access to the outdoors and pasture, and spend limited time in confinement. Organic confinement systems are typically less crowded than conventional confinement operations. At the core of many organic livestock systems is the grasing system. Animals forage on pastures for their own feed while spreading manure, yielding energy and labour savings, and reduced equipment costs. Composed of legumes, grasses and other broadleaf species, pastures provide multiple benefits for the soil and ecosystem as well.

Nick Maravell, who has been farming organically since 1979, branched into beef not long after he increased his Buckeystown, Md., operation to 165 acres in 1997. "When I expanded, it just made sense to become more diversified," said Maravell, who was already growing organic forages, hay and grains. "When you have animals, you complete the cycle of feeding the vegetative matter through the rumen, and it comes out the other end to fertilise the soil."

Maravell, who pastures a small herd of Black Angus on 16 acres, also likes the additional flexibility. "The cattle can be used in different ways," he said, "by sending in the cows to cull the crops—rye, barley, alfalfa, or even soybeans, when green—if you decide you don't want to harvest them." Although pastured animals are not necessarily organic, and some organic producers do not extensively pasture their livestock, organic growers can capitalise on the benefits of grass-raised animals by marketing their product as "grass-fed" as well as organic.

Maravell feeds no grain to his cattle and grows all his own grass and hay on his farm. The lack of grain means that his animals do not have high intra-muscular fat -which fetches a higher USDA rating than the lower fat meat— "but," he said, "my customers want this beef because it is grass-fed and organic. They've seen articles documenting the benefits, and they don't want antibiotics, hormones or pesticides, all of which you eliminate when you raise animals organically."

Maravell, who sells direct to his customers, added that another benefit of pasture-raised beef are the higher levels of conjugated linoleic acid (CLA) and omega-3 fatty acids. "There is good evidence that conjugated linoleic acid can prevent cancer in animals and may protect against heart disease and diabetes and obesity in humans; while omega-3 fatty acids have the potential to decrease the risks of cancer and cardiovascular disease," said Donald Beitz, a Professor of Animal Science and Biochemistry at Iowa State University.

Beitz, who collaborated on a SARE-funded study in Iowa on the effects of pasturing animals, added that CLA levels can be 4 to 6 times higher in the milk of cattle who feed on fresh pasture versus cattle who eat stored feed such as silage hay and grain. Karsten, the Penn State researcher, also found that the eggs of pastured poultry had three times the amount of omega-3 fatty acids and higher levels of vitamins A and E than did the eggs of conventionally fed poultry.

Managing good health for organic livestock presents challenges since no antibiotics or hormones may be used. In general, the focus on animal health is preventive, and the incidences of diseases typical to animals in confinement can be reduced or mitigated by pasturing the animals, focusing on good nutrition and providing preventative care. According to Karsten, there is evidence that the health of pastured animals is better. "Since the animals are not standing on cement, they have fewer foot and leg problems, the fresh air reduces respiratory problems and the incidence of mastitis is decreased since the animals are not lying in their own manure in the barn."

Dairy farmers also successfully manage their herds without the use of the standard conventional treatments. "Some people believe that you must use penicillin or manage dry cow treatment with antibiotics, but organic farmers don't add anything for dry cow treatment," said Lisa McCrory, the dairy technical assistance coordinator for the Northeast Organic Farming Association of Vermont (NOFA-VT). "If a cow does get mastitis, homeopathic methods and colostrum products work well."

Focusing on animal nutrition through high quality feed and good soils also goes a long way toward reducing stress, illness and the need to treat animals for medical problems, added McCrory. Hubert Karreman, a veterinarian in Lancaster County, Pa., who has been treating organic dairy cows for almost 10 years, agreed that the incidence of many diseases is lower in organic and grased herds. "But," he added, just because you're using organic management does not mean you won't have health problems." He sees more pasture bloat and hoof punctures and abscesses in grased animals, but also believes that preventive strategies such as probiotics, and homeopathy and botanical medicines can be used very successfully to manage and treat organic herds.

ECONOMICS OF ORGANIC PRODUCTION

Organic farmers are often the first to admit that as they were transitioning to organic systems, their yields declined. Many studies have shown that, initially, a decline in yields occurs during the conversion to organic production. However, once the transition period has passed -usually in three to five years—organic crop yields often rebound to within 90 to 95 percent of conventional yields, according to an Organic Farming Research Foundation review of comparative studies.

Perhaps even more important, once the farming system has been certified, price premiums for organic crops, added to the reduced production costs, help boost profitability. For many organic farmers, equivalent yields are not necessarily the goal. "High yields are not always connected to profitability," said Wende Elliott. On her farm, she expects a 37.5 percent operating profit margin, largely due to lower input costs and a premium price for organic poultry, hay and row crops.

Jeff Moyer, farm manager at The Rodale Institute in Kutztown, Pa., explains in a fact sheet published by the Institute how organic farming makes good economic sense. In 2001, his organic corn and soybean yields were only 90 percent of conventional yields, yet the organic corn fetched $4.70 bushel compared to $2.10 for conventional. The soybean price disparity was even larger—conventional soybeans went for $3.80 per bushel, while organic livestock feed beans brought $10.50 per bushel and organic food grade beans $15 a bushel.

Ed Fry, who farms 400 acres of grain and has 240 milk cows in Chestertown, Md., points out in a marketing fact sheet from Rodale that while his corn yields were comparable in 2000, his total production costs were lower for organic corn—$1.79/bushel versus $2.23 for conventional. The labour per acre was higher in his organic corn, but because the organic corn fetched $4 a bushel versus $2.50 for the conventional, he didn't need to farm as many acres for the same amount of profit. In organic dairy operations, a similar principle of reduced production and higher profits applies. When Vince Foy and Debbie Yonkers of North Danville, Vt., converted their 70 Jersey cows to organic, their milk production decreased by 10 to 15 percent, but their gross income increased from $125,000 to $165,000.

Moreover, they cut their debt-to-cow ratio in half. In fact, said Lisa McCrory, dairy technical assistance coordinator for the Northeast Organic Farming Association of Vermont (NOFA-VT), "organic dairy producers almost always reduce their production numbers, due to management changes such as feeding the animals less grain." And even though the price of organic grain is higher, other costs such as veterinary bills, fertiliser and labour decrease, improving net income.

A statewide study conducted in Vermont by the Northeast Organic Dairy Producers Alliance showed that although milk production was lower in the organic systems, the organic producers received an average net return of $477 per cow per year compared to the conventional average of $255 per cow.

"Looking beyond production and making decisions based on profitability and the bottom line makes good business sense," McCrory said. While more research is needed on the economics of transition, the longterm economic viability of established organic systems is quite positive. A 1999 Wallace Institute review of six midwestern land-grant university studies found:

— Organic grain and soybean production systems are "competitive with conventional production systems."

In fact, with current market premiums, producers of organic grain and soybeans earn higher profits than conventional growers.

— Without a price premium for organic crops, half of organic systems were still more profitable than the conventional systems. Those systems less profitable than conventional quickly surpassed the conventional systems when organic premiums were figured in.

In cases where organic systems were more profitable without price premiums, it was generally due to lower production costs, higher net returns due to the types of crops in the organic systems, and better performance of the organic systems under drought conditions or in drier areas. Production costs tend to be lower in established organic systems because of reduced input costs. One exception to this, perhaps, is labour.

Organic farming systems are often more labour intensive because of increased time spent managing weeds and monitoring pests. Labour costs, however, can be measured in different ways. "If a farmer views his/her time spent on the farm in terms of its opportunity costs, e.g., what he or she could be earning off the farm, labour costs for organic farming are higher than conventional," said Jim Hanson, extension economist in the Department of Agricultural and Resource Economics at the University of Maryland. "However," he added, "for those farmers who don't view off-farm income as an alternative source of income the labour costs between the two systems are similar."

In a forthcoming study to be published by Hanson, he found that family labour was about 30 to 40 percent higher in an organic mid-Atlantic grain operation than in a conventional one, but hired costs were equivalent between the two systems. Production costs also vary by region, climate and production system. For example in humid areas, pest and weed control measures can raise costs.

A recent study in a corn-soybean system in Iowa found costs of conventional production were only slightly higher than organic. The organic farms had lower fertiliser and pesticide costs, but higher seed and machinery costs. However, in a SARE-funded project that compared organic and conventional apple production across California, Sean Swezy, formerly a researcher at the University of California and now director of UC-SAREP, found production costs of organic apples 10 to 25 percent higher than conventional ones in the coastal fresh market systems due to labour and material costs. However, statewide, the organic systems were determined to be commercially profitable.

Finally, in a SARE-funded potato study in Idaho comparing 18 conventional and organic farms, the average material costs were lower in the organic and the labour costs higher, but overall there was no significant difference in fixed and variable costs. Organic livestock systems often cost less, thus can be a viable option for beginning farmers or those who have trouble raising capital, because those systems do not require elaborate or expensive housing. Poultry, for example, can be raised on pasture using inexpensive, easy-to-build structures.

Risk Management

On the surface, Organic farming may seem to be riskier: Organic farmers are limited to a much smaller range of pest control materials, are learning a new management system and may be subject to pest and weed outbreaks during what is known as the "ecological" transition period. Fred Kirschenmann, director of the Leopold Center in Iowa and a long-time North Dakota organic farmer, believes that the nature of organic farming and its reliance on crop rotation provides protection against vulnerability.

"Diversity spreads out risks and vulnerabilities," he explains in the introduction to Cereal-Legume Cropping Systems. "If you have one or two crops, you are vulnerable to market and natural adversities of that narrow band of crops. That makes farming a high-risk venture. As diversity on the farm increases, growing risks get spread out and market opportunities increase." Substantial expansion in one product could cause prices to drop, so premium prices are not always a guarantee.

Most research, however, seems to indicate that organic row crop and small grain rotations can be competitive, even without the price premium. Diversifying the production system is paramount for economic security. "If you're going with contracts, you need to plan very carefully because organic premiums fluctuate a lot," said Steve Temple, the UC-Davis researcher. "It's OK if you have a high-value crop in the rotation but it's more important that producers count on more modest premiums on all crops than simply depending on just one high-value premium. The market for high-value crops can get saturated and the prices can decrease."

A six-year study in northeastern South Dakota comparing organic and conventional systems found much less variability in net income over all costs, except for management, in the organic system. The net returns varied by \$16/acre in the organic system versus \$31/acre in the conventional system and there were no negative net returns in the organic system. Economic effects also vary across different enterprises and regions.

For example, an organic apple enterprise might be appropriate for western areas with low disease and pest pressure, but is much tougher in eastern areas, where production costs are too high to price apples competitively. There is less yield risk in organic grains in dryland areas. Finally, enterprises based on direct marketing are less likely to succeed if the grower is far from a major population center. During transition, risks associated with extra costs such as re-tooling, purchasing additional equipment, extra storage requirements and additional labour can be substantial.

Wende Elliott and Joe Rude helped form a marketing co-op to share some of those risks. "If we had tried to be a one-man show instead of joining a cooperative group," said Elliott, "it would have hurt us because we all would have been competing against each other based on price." Co-op members share the labour associated with marketing, such as promotion and sales. And by working together on quality control, codification of nutrition and genetics, the co-op members each gained access to bigger markets.

Moreover, the growers had more time for their individual operations. As start-up farmers, Elliott and Rude shared, rather than purchased equipment, whenever possible. "Having a strategy for managing production, marketing and financial risks is important during the three-year period it takes to transition to organic," said Sharon Hestvik, small farm coordinator for USDA's Risk Management Agency.

"One risk management strategy for producers transitioning into organic is to keep good records." RMA provides risk management tools and crop insurance coverage on over 100 crops for both organic producers and those transitioning into organic farming practices. RMA requires several documents, including:

— Records from the certifying agency showing the locations of fields that are transitional, certified organic, buffer zone acreage and conventional;
— A copy of written documentation from a certifying agent showing an organic plan for the acreage;
— Records of the types of crops grown, yields and whether the crop is irrigated or non-irrigated; and
— The dates that the crops were planted.

As part of the organic production learning curve, costs may go up. Farmer experience and research clearly shows the transition loss can be reduced, however, by taking the time to learn what you're doing and convert a little at a time. At Taconic End Farm in Vermont, where profits climbed 40 percent during the transition, Annie Claghorn's and Caitlin Fox's only change in management was the purchase of organic grain, since their cows were already pastured.

John Vollmer, the N.C. strawberry farmer, attributes part of his successful transition to the fact that prior to converting he had been using organic soil management techniques for two years. Some states, such as Minnesota and Iowa, also provide technical and financial assistance for growers in transition.

Marketing

The methods that organic farmers use to market their products are as diverse as the types of organic farming systems proliferating across the country. According to the OFRF survey, about 80 percent of organic farmers market through wholesalers, 13 percent sell directly to the consumer at farm stands, farmers markets and local restaurants, and the remainder market direct to retail outlets or stores. One of the hardest challenges for farmers converting to organic is learning how to market products before growing them.

Yet, many organic farmers become very successful marketers, particularly those farmers who direct market. They work hard to build relationships with their customers, and rely upon creativity, education and development of alternative outlets, such as alliances, cooperatives and other resource-pooling ventures.

"Many of the most successful growers are the best marketers," said extension agent Brad Brummond. "Longterm growers have good relationships and are top-notch marketers. They spend time figuring out what the market wants and they negotiate prices, because in organic systems farmers can really affect the final price by the relationships they develop with buyers."

Farmers markets are an excellent direct market venue for organic farmers. Between 1994 and 2002, the number of U.S. farmers markets grew 79 percent, reflecting the expanding market base. "You may not see a big premium at a farmers market, but you get more customers, people get to know you, and most markets welcome organic farmers," said Dan Nagengast, executive director of the Kansas Rural Center and an organic farmer.

Grower Alliances: Pooling resources can be invaluable. Organic growers in North Dakota lacked lucrative markets for their fresh produce, meat, grain and value-added products. Ben Larson, a researcher and organic farmer, contacted the Organic Alliance in St. Paul to develop a marketing strategy, consumer education information, and a media plan. Larson also went directly to large grocery stores to introduce available product. He then coordinated all the interested local growers so the stores would only need to make one call a week for their order.

As part of the plan to help educate consumers, Larson provided stores with "point of sale" materials and advertised on public radio and in newspapers to promote organic foods. He also started a new farmers market to focus on locally grown foods. As a result, sales increased at the farmers market and the grocery stores—and their success selling organic potatoes encouraged the grocers to try other products. "We were trying to reach the larger segment of the population who will choose organics if they're available in the grocery store," said Larson.

Restaurants: Many organic farmers direct market to high-end restaurants and farmers markets. A New York chef, quoted in a cover story on organic agriculture in a fall

2002 Newsweek said, "When people taste asparagus or string beans grown in richly composted soil, they can't get over the depth and vibrancy of the flavour."

The farm-restaurant relationship has worked well for Urban Oaks Farm in New Britain, Conn. "Even if you grow the best tomato in the world, if you can't sell it, it isn't going to work," said Urban Oaks co-manager Tony Norris, who grows greens, herbs, tomatoes and eggplant, among other vegetables. Norris sells much of his organic produce to Hartford restaurant chefs based on relationships he built with care. He advises farmers to arrange an appointment with a sympathetic chef, and bring a sample of products, a price list and clear billing and delivery system. "You have to think it through," he said. Norris considers himself a "consultant" to the chefs he supplies, but "if you're not comfortable doing that, maybe a partner or spouse can do it."

Marketing Companies: Organic farmers Richard and Peggy Sechrist of Fredericksburg, Texas, who have a 50-head herd of beef cattle and raise 750-1000 pastured chickens per month, formed a company specifically for marketing purposes. When they found it too difficult to reach the volume they needed to turn a profit in direct sales, they developed a label to differentiate their products and fetch a premium.

Under this label, they now sell their own products and those of neighbouring ranchers raising organic meat. Sales go to an initial customer base of about 750 built through mail order, farmers markets, booths at fairs, and small health food stores, but new purchasers also find them. Their financial success comes from the strong market for chicken and relationships with food distributors. But they also work constantly to educate consumers about their product, how their meat was raised and the issues around organic farming.

TRANSITION TO ORGANIC PRODUCTION

Converting to organic production is not a decision to take lightly. Organic farmers must learn how to work with nature to solve problems, such as adapting crop rotations to improve soil fertility, manage weeds and control pests rather than simply substituting accepted materials for prohibited ones.

Farmers considering a transition to organic farming should think about the following questions, drafted by the Ohio Ecological Food and Farming Association (OEFFA):

— Do you enjoy walking your fields on a regular basis?
— Can you distinguish pests from beneficial insects?
— Are you curious about why things happen on your farm?
— Can you tolerate a field that is not weed free?
— Do you have the patience to trade short-term economic returns for longer-term "ecological" credits while building soil health?

Farmers converting to organic purely to improve profits often fail because they do not consider the huge range of economic, social and production changes that must occur. The transition period can be particularly stressful because of the need to develop and implement new management skills. In fact, you must be prepared to survive a short-term financial loss if yields drop and costs increase during this period.

Other considerations, posed by OEFFA and others, include:

— How will the transition period, where yields sometimes decrease and price premiums are not yet available, impact your family?
— How will social stigma and negative peer pressure from other farmers impact you? Lydia Poulsen, the Utah cattle and grain farmer, said that when she

converted a decade ago, people couldn't understand what she was doing, or why. Now, people seem more accepting and interested.

— What resources are available? Consider labour, borrowing capability, knowledge base of local extension and information exchange regarding organic production. When the organic dairy industry was expanding in Vermont, SARE-funded researchers designed a study to explore business, crop and animal management on dairy farms. One the most important findings of the study, said researcher Lisa McCrory, was "how all the growers benefited from what they could share with each other, the time spent brainstorming and sharing through on-farm demonstrations and conferences, and exchanging pasture management skills."

— How will you develop the new types of relationships required to market organic products?

Some farmers view the transition period as an investment in education. During this time, when some growers experience declining profits, remember you are not only learning new skills but also are building what some economists call "natural capital." This refers to improved physical characteristics of soil and plants, such as better soil water infiltration, increased microbial populations, more natural predators, and better control of the weeds. Like investing in a new stock, there may not be short-term profits, but in the long run, you are setting the stage for the sustainability of your land and farm.

After deciding to take the plunge into organic production, consider the following strategies:

— Identify the closest certification organisation and start collecting information about how to come into compliance.

— Make contacts. Attend meetings of organic and other transitional farmers, collect books and other resources and find extension agents and other educators who are knowledgeable about organic production and transition strategies.

— Experiment with a systems approach that will work on your farm or ranch. Focus on prevention strategies and treating the causes of problems rather than specific problems themselves.

— Develop marketing strategies for your organic products.

In the planning phase, evaluate the strengths and weaknesses of the farm. How will you work with the natural system you have? Are you ready for a transition? For example:

— What are your most valuable natural resources?

— Does topography work for or against you?

— What kind of pest pressure do you experience?

— How healthy are the soils?

Soil health is extremely important because you will no longer rely on external inputs, but depend instead on the activity and capacity of the soil. "I knew that my soils were basically sand with a little bit of nutrients and that everything was burned out," said John Vollmer about why he took two years to even start the transition. When breaking new ground or exploring new enterprises," Spend the money to get good soil tests done so you know what amendments you'll need," said Tony Norris, an organic vegetable grower in New Britain, Conn. Think about pest control. Biological pest control is complex, involving complicated interactions among crop rotations, intercropping combinations, planting schedules and beneficial habitats. What strategies or systems are already in place?

Fred kirschenmann, a north dakota farmet and a long time advocate of sustainable farming systems, gives this advice in the foreword to Cereal-Legume Cropping

Systems, a book for farmers who are exploring switching to sustainable production methods.

"In switching from one system to another, it is extremely important to remember that what one is doing is switching systems and not changing technologies. Many farmers have experienced severe unnecessary losses in making the switch because they failed to appreciate the difference.

"Making this switch is not simply a matter of substituting green manure for synthetic fertilisers or substituting organic fertilisers for conventional ones. Nor is it primarily substituting botanical pesticides for 'toxic' ones. Making the switch is a matter of slowly backing out of one system of farming and slowly introducing another system of farming. "Effective use of on-site nutrient cycles as the primary source of fertility requires good organic matter, a crumbly soil structure, minimum compaction and high levels of biological activity.

"Effective pest control with reduced or no pesticide use also takes time. It takes time to break the pest cycles that have become established under monocrop management. Good pest control with fewer off-farm inputs requires a thorough understanding of the natural cycles of the weeds, insect pests, and diseases that have established themselves on your farm and then determining what crops and practices are most effective in interrupting those cycles."

Nancy Creamer, a leader of the North Carolina State University organic transition experiment comparing a range of organic systems, has benefited from the cumulative knowledge gained from the last two decades of organic farming research. Based on past research results, Creamer and her colleagues started their rotation with soybeans instead of corn and applied principles of organic weed management to achieve relatively weed-free fields. "It's possible that it's easier to farm organically now, given the cumulative experience that has been gained by farmers over the years," said Creamer.

By designing research based on results from earlier studies, Creamer and others have shown that is possible to make the transition with minimal production losses. By preparing the land, building soil, focusing on the right crops and rotation, and not putting too much acreage or too many animals into production, farmers can minimise what has come to be known as the "transition effect." Creamer's study, for example, showed that with good weed management, soybean yields can equal those of conventional beans during the first year of a transition.

Management Strategies

Rotations: Start by deciding how to build your rotation, as this is the most important management tool in an organic system. Your biggest challenges likely will be weeds and nitrogen fertility, so think carefully about how to balance those constraints with maintaining a high-value crop. Grass/legume mixtures provide good cover and supply nitrogen, but if your soils are low in organic matter, you may need to incorporate the mixtures instead of cutting them for hay.

Research cash crop alternatives that will help steady your bottom line. At the Rodale Institute, a five-year rotation was cut to three years to get more high-value cash crops and to allow the grass/ legume mixture to be used for improving soil fertility. Mary Howell Martens, an organic farmer from Penn Yan, N.Y., recommends striving for balance between maintaining soil health and producing economically profitable yields.

"One of the problems we've seen [is farmers] putting the whole farm into the most profitable organic crop every year. With no rotation, the yields go to pieces," she said, when interviewed for the SAREfunded educational video,

"Organic Grain: Another Way." "The crop mix that we have developed maintains our soil health and yields and also gives us a fairly profitable operation. But the overriding consideration is: What does this field need? We do put in some crops that are not particularly profitable, but then they're being rotated with the crops that make us the best income."

Begin with cash crops that require less nitrogen and can be effectively managed to control weeds. Many studies have shown that, with proper weed management, soybeans can be planted in the first year of a transition with no declines in yields. Corn, on the other hand, is not a good transition crop because it requires a lot of nitrogen and more weed management. However, once the system is established, organic corn can be grown quite successfully, said Kathleen Delate, an Iowa State University organic researcher. Her studies show that within three years of the transition, organic corn can produce as well as conventional corn.

Avoid consecutive years of row crops to prevent weed outbreaks and maintain system productivity. University of Minnesota researchers and a dozen or so Minnesota farmers collaborating on a SARE project found that avoiding consecutive years of row cropping during the conversion to organic production could prevent weed outbreaks and maintain system productivity.

The experiment, which examined the effect of crop sequence, showed that in the third year of production, the organic systems where corn or soybeans had been preceded by one to two years of small grain/forage legume versus a row crop had lower weed incidence and higher yields by the third year. This suggests that you should avoid planting two row crops in sequence in the transition.

Other considerations for the rotation:

— Does the rotation match the crop needs for fertility? Try

to have crops with differing root depths so they can access different nutrient zones.

— Does the rotation have sufficient diversity so that risks will be minimized?

— Does it provide weed control?

Take advantage of mixtures and niches such as the combination of sorghum-sudangrass/lablab/ cowpea planted in early summer following tomatoes and preceding corn in a California experiment. "With a nitrophilic crop like corn following tomatoes, it's important to have a mixture that's building the soil," said Steve Temple, one of the researchers. The mixture, he continued, is designed for multiple functions:

— The sorghum-sudangrass germinates quickly in the heat and takes up residual nitrogen. It also shades out late summer weeds.

— The cowpea, which fixes its own nitrogen, grows productively alongside the sorghum-sudangrass.

— When the sorghum-sudangrass dies back in early fall, the lablab, also a legume, emerges, so that by December, when all three species have winter-killed, the dense mass of cover contains fixed and recycled nitrogen.

— With all the nitrogen accumulated and fixed in the fali, and stored in the vegetative biomass over winter, you don't have to wait for spring growth of the cover crops. The dense mixture can be plowed under and the crops planted as soon as the ground is ready.

Soils: Building soil organic matter and improving soil quality is often cited as the most critical step for a successful conversion to organic farming.

— It may take three to five years for the soil to improve, depending on the condition of the soil, so start adding manure or composts and finding other sources of organic amendments as soon as possible.

— Balance production, soil building and conservation. Good crop rotations that include cover crops and animal wastes help build soil organic matter.

— Do research and start experimenting. Reading a book on compost, "made me realise I should just do things instead of figuring out the technical parts," said John Vollmer.

Weeds: One of your biggest transition challenges will be weed management. New studies are showing, however, that with careful management, weeds can be controlled effectively during transition:

— Careful weed control was one reason that an Iowa study found no yield reduction in soybeans, and loss in corn only for the first year of a transition trial. "We attribute our results to high managerial experience in producing diverse crops and accurately operating various implements in organic systems," said project leader Kathleen Delate.

— A SARE study in Minnesota found that a crop of buckwheat harvested for seed was effective at smothering Canada thistle, in both the immediate and the subsequent crops—winter rye and soybeans. The research also showed the economic impact of good weed management; every bushel of soybeans not lost to weeds increased profitability $12 to $18 an acre. Moreover, replacing four rotary hoeings or harrowings with two well-timed ones for Canada thistle reduced costs by $3 to $5 an acre.

A change in attitude toward weed management is also critical, said John Hall, a Maryland extension educator who co-created the "Organic Grain: Another Way" video. "We think a field has to be weed-less to be productive. What we're seeing by those in transition, though, is that we can tolerate weeds. We just have to know where the threshold is and be willing to accept that."

John Teasdale, a weed scientist with USDA's Agricultural Research Service who has studied non-chemical methods to manage weeds, focuses on controlling seeds. "One aspect of trying to grow crops without herbicides is to control the weed seed population and keep it as low as possible," he said. "It is important not only to try to control weeds and prevent yield losses, but also to prevent those weeds from going to seed and building up a soil seed bank."

Other strategies to consider:

- Identify weed problems before they start, addressing them in crop rotations.
- Distinguish between annual and perennial weeds, as well as those that spread by rhizomes and seeds, to develop effective management.
- Plant higher crop densities to block weed germination.
- Shift between warm- and cool-season crops in your rotation to disrupt the life cycle of various weeds and reduce competition.
- Include crops that have natural weed inhibitors like rye and sorghum.
- Plant crops that can be sown late in the season and easily cultivated. Switching to transplants in horticultural crops can provide a jump on the season, and allows more soil to be thrown up around the plant without causing damage.

Insect Pests: Plan your rotation and soil-building strategies to manage insects and diseases. Be aware that elimination of pesticides can lead to temporary outbreaks of pests. Before starting the transition, "minimise pesticide applications, and use pesticides with the least impact on natural enemies," said Abby Seaman, an extension educator with the New York State IPM program and Cornell University, who received a SARE grant to investigate the relationship between management practices and pest populations. "This will make the transition less jarring." Seaman also recommends:

— Push the envelope with IPM practices, such as scouting and setting thresholds for pest populations. Gain experience spotting natural predators in the field.
— Become familiar with acceptable management materials and start trying them.
— Build soil organic matter to reduce disease pressure. Livestock. Many farmers and ranchers who are already using pasture-based systems to raise their animals don't find the transition difficult.

Beef producers in Nebraska are entering the organic market by using the pasture-based systems they've perfected over the last few years, said Martin Kleinschmit, sustainable agriculture specialist with the Center for Rural Affairs in Walthill, Neb. "The transition is easy," Kleinschmit said, although certain rules must be met as you convert to organic production.

Ranchers need to provide buffer areas—25-foot fenced setbacks from conventional neighbours -on their pastures and keep animals out of streams. Similarly, deep-bedded systems, such as those used in "hoop" barns for hogs, create a jumping-off point for would-be organic pork producers. With help from a SARE grant, Minnesota pork producer Dave Serfling decided to create a deep-bedded system for his 170 hogs. He converted an old, two-story barn into a straw-based system, an efficient way to generate heat through the animals' body warmth and manure, which composts in the straw. Not only does Serfling save on heating bills, but he also avoids manure storage concerns because the manure-straw mixture creates an ideal crop fertiliser.

"They can even stay warm on days when we record 30 degrees below zero," Serfling said. The hogs are able to grow in a group, exercising and interacting in a herd setting rather than living in individual crates. Serfling sells his organic pork to an upscale food retailer that established animal welfare guidelines, including a no-crate rule, and supplies restaurants and premier retail stores. Serfling receives at least a 6-centper-pound premium.

Dairy farmers are well positioned for transitioning to organic if they use pasture as a major feed source and don't over-push their cows for production or utilise many antibiotics or hormones, said Lisa McCrory. Even if transitional farmers don't follow all these strategies, they can start one at a time by trying alternative dry cow therapies, eliminating prohibited materials such as hormones and antibiotics, and getting the cows grazing. Pastured cows tend to need less medical treatment and antibiotics because of access to fresh air and exercise, so pasturing is an excellent way to begin a transition.

Depending on a farmer's comfort level with grasing, it can take anywhere from one season to five years to learn to take full advantage of a pasture. Many organic dairy farmers have reduced production goals yet still turn a higher profit than conventional operators because input costs such as veterinary bills, drugs and feed decrease. Although the land to produce the organic grain must be managed according to organic standards for three years, the cows only need to be managed organically for one year, so some operators transitioning to organic sell off the milkers and keep the young stock for transition.

Heifers eat very little grain, and most will be in compliance since they haven't received antibiotics, or drugs for dry-cow treatment. Poultry farmers who have adopted outdoor, minimal confinement systems also have a similarly easy transition to organic systems. The small but growing practice of raising broilers, layers and turkeys in pasture-based systems lends itself to organic certification because it meets two of the requirements of the national rule for organic meat—outdoor access for livestock and elimination of antibiotics in feed.

Most alternative poultry producers already avoid antibiotics, saying birds not crowded together in confinement systems experience fewer infections. Producers still need to watch for diseases and weather -related stress. To control such incidence, consider:

— Moving the birds frequently, allowing pathogens to die off when their food source is removed.

— Cleaning pens and brooders regularly between flocks.

REFERENCES

Dorothy Myers and Suer Stolon (ed.), *Organic cotton from field to final product*, Intermediate Technology Publication, 1999, 250 p.

Louis C. Kervran, *Biological Transmutations*, Published by Swan House Publishing Company, Bingnampton, N.Y. 1972.

Pfeiller E., *Biodynamic Farming and Gardening*, Published by Pauma Valley Company, California, 1973.

Thampan P.K., *Organic Agriculture*, Peekay Tree Crops Development Foundation, Cochin, Kerala, 1995, 354 p.

5

Progress of Organic Farming

The first conference of NGOs on organic farming in India was organised by the Association for Propagation of Indigenous Genetic Resources (APIGR) in October 1984 at Wardha. Several other meetings on organic farming were held at different places in the country towards the end of 1980s. Here, mention must be made of the Bordi Conference in Maharashtra, the state which was the focal point for the organic farming movement in India.

The Rajasthan College of Agriculture with the support of the state government organised a meeting on organic agriculture in 1992. The United Planters' Association of South India (UPASI) organised two national level conferences on organic farming in 1993 and 1995. ARISE (Agricultural Renewal in India for a Sustainable Environment) is a major organisation in the country engaged in the promotion of organic farming. ARISE was founded in 1995 at a national conference of organic farming held at Auroville.

ARISE comprises of a supporting network of regional groups aiming at sustainable environment by protecting biodiversity and promoting organic agricultural practices. The selection of Auroville for the conference was apt as it housed the Arabindo Ashram and the pioneering work under its auspices on building technology, alternative energy research, wasteland development, afforestation and organic agriculture. By 1980, three groups of Indians had taken to organic farming. The first one consisted of urban educated technocrats for peripheral interest, which did not last long. Educated farmers consisted of the second group whose farming practices were based on scientific knowledge.

The third group practiced organic farming through trial and error. The successful organic farmers in India are those who have access to sufficient natural resources like, water and other organic inputs mostly on their own farms. These farms produce crops like sugarcane, areca, cocoa, coconut, pepper and spices. Many of them have shown that switch over to organic farming do not affect yields and income and more importantly, knowledge/ expertise is available for successful adoption of organic farming in the country.

The International Federation of Organic Agriculture Movements (IFOAM) estimates that an area of about 41,000 hectares in India is under organic farming representing about 0.17 per cent of the world organic acreage. It also reveals that the percentage of organic area to the total cultivated area comes to only about 0.03 per cent and the total number of farms comes to about 5,661. But, a comparison of our 41,000 ha to Australia (10.5 million ha), Argentina (3.19 million ha], Italy (1.83 million ha), and USA (0.95 million ha) clearly indicates that organic farming in India has to go very far even to catch up with that of the leading nations of the world.

Non Governmental Organisations (NGOs) are spearheading organic farming in India. A report in 2002 indicates that about 14,000 tonnes of organic products have been raised in India. They include tea, coffee, rice, wheat, pulses, fruits, spices and vegetables. India exports organic agricultural produces to European Union, USA, Canada, Saudi Arabia, UAE, Japan, Singapore and Australia, among others. The International Conference on "Indian Organic

Products-Global Markets" at the end of 2002 was the first to be held in India.

IFOAM predicts that India and China have great potential to be organic farm produce exporters in the future. An important event in the history of the modern nascent organic farming in India was the unveiling of the National Programme for Organic Production (NPOP) on 8"^ May, 2000 and the subsequent Accreditation and Certification Programme on P' October, 2001. The logo "India Organic" was released on 26'*' July 2002 to support the NPOP.

An important progress towards organic agriculture made by India is the increasing awareness of the ill effects of the modern farming system, which the country adopted about 35 years ago. The threat poised by the conventional food products to the human health and the damage done to the ecology are being viewed seriously. Efforts are made to produce healthy foods and the demand for them is increasing. The importance of the marketing of the organic products is highlighted for the promotion of organic agriculture. Several individuals and associations have taken to organic farming and organic products are available in the large cities to a very limited extent.

PRODUCTION AND EXPORTS

The aggregate production of organic agriculture came to about 14,000 tonnes during 2002 and the exports amounted to 11,925 tonnes.

Regulations

The most important step towards organic farming taken by the government was to draw a regulatory framework. It is true that the initiatives by the government to introduce organic farming by laying down regulations came belatedly as many countries have already done this kind of basic work decades ago. The implementation of NPOP is ensured by the formulation of the National Accreditation Policy and Programme (NAPP).

The regulations make it mandatory that all organic certification bodies should be accredited by an Accreditation Agency. The international certification agencies operating in India even prior to these regulations will also have to get accreditation under the new dispensation. The regulations lay down the institutional arrangements for implementing the national programme for organic production.

The NPOP is administered, monitored and implemented for the benefit of farmers, processors, traders and consumers. It envisages a three tiered organisation under the overall guidance of the Union Government with the Department of Commerce, Ministry of Commerce and Industry as the nodal agency. Policy making and declaration of the standards for organic products, recognition of organic standards of other nations, efforts to get our standards recognised by others and coordination with other arms of the government for the successful management of the organic agriculture are the major functions entrusted to the ministry.

The agencies accredited are the Agricultural and Processed Food Products Export Development Authority (APEDA), Coffee Board, Tea Board and the Spices Board. The regulations cover exports, imports and the domestic trade of the organic products. But the government regulations are applicable to only the exports. So, an organic farm product can be exported only if it is certified by a certification body accredited for the purpose.

The categories of products covered under accreditation are organic crop production, organic animal production, organic processing operations, wild products and forestry. A national level steering committee is functioning as the apex advisory body for assisting the government to promote organic farming in the country. This body consists of representative's form the Ministries of Agriculture, Food Processing Industries, Forests and Environment, Science and Technology, Rural Development and Commerce. Organic production requires certification after periodic inspections in order to ensure that all prescribed practices are followed.

The inspection and certification are done by the agencies accredited to the Accrediting Agencies dealing with the commodity. Inspection and certification agencies can be government departments, NGOs, trade or consumer or producer organisations. Such agencies should be registered bodies, with managements in position, declaring the persons who shall be held responsible for any miscarriage of certification and having proof of adequate field staff to undertake periodic inspections.

The continued accreditation of such bodies is dependent upon their record of fidelity to the principles of organic production. They are authorised to award certificates after due satisfaction that practices conformed to those enunciated by the Accrediting Agency in relation to the item concerned. The charges levied by the certifying agencies are fixed by the Accrediting Agencies.

Research in Organic Agriculture

A National Institute for Organic Farming has been established to spearhead research in organic agriculture. The government of India constituted task force had also recommended the initiation of the postgraduate level courses in organic farming. The Morarka Foundation and Maharana Pratap University of Agriculture and Technology (MPUAT), Rajasthan have collaborated in the design and implementation of such a programme.

PROJECTS AND INITIATIVES

Several projects and initiatives to promote organic farming

in the country have begun at the behest of individuals and institutions. The following are only a few of such efforts the details which could be available. A project aided by the World Bank to empower the rural communities in the country to grow organic products for exports had come up in 2002. The programme aims at the improvement and promotion of organic production of spices, certification and export of black pepper, white pepper, ginger, turmeric, cardamom, clove, nutmeg and herbals like rosemary, thyme, oregano and parsley.

The implementation of the progamme is done by the NGOs, and Idukki and Waynad districts of Kerala, Nilgiri district of Tamil Nadu and Kandhamal district of Orissa are the areas selected for the purpose. Imparting training to both the JNTGOs and the farmers on organic production methods, basic standards required, documentation, inspection and certification is a major objective of the programme. The assistance to NGOs includes among others computer hardware and software especially for market promotion of their produces.

An initiative for the spread of organic farming by various stakeholders in the Indian organic agriculture sector is their coming together to constitute an apex body for providing centralised services and expertise for the increasing number of organic farmers in the country. Initial steps for setting up the Indian Competence Centre for Organic Agriculture (ICCOA) has been taken at a meeting held at the National Academy for Agricultural Sciences in New Delhi in 2003 under the joint auspices of INDOCERT, a Kerala based organic certification agency and the Swiss based FiBL (Research Institute of Organic Agriculture).

The decision was the outcome of an Indian team's visit to Switzerland in 2003 to study the structure of Swiss organic farming. The meeting elected a nine member promoter board of directors to facilitate the setting up of ICCOA, which would be registered as a charitable society.

The Centre would strive to strengthen and supplement the efforts of the state governments in promoting the development of organic agriculture in the country.

In Haryana, an enterprising farmer who began farming on his 16 acre land in the Sonepat district in 1971 could establish an organic farm on 108 acres, raising vegetables and other crops. It appears that his success is mainly attributable to the efforts made by him to market the products. NGOs functioning in the neighbourhood of Delhi buy the produces like rice, wheat, pulses and vegetables from his farm at a premium price of 30 to 50 per cent.

An exporting firm at a premium of 20 to 30 per cent buys his basmathi rice. Almost 70 per cent of his farm production is sold through advance agreements/contracts. Presently he and his friends are working with the resident associations in Delhi to market their organic products at a premium of about 25 per cent.

In Rajasthan, the Morarka Foundation, established in 1995, promotes sustainable agriculture. It has about 10,000 partners producing vermi-compost and the Foundation is said to be the single largest producer of this organic input in Asia. It encourages the production of bio-pesticides and supports procurement of certified organic products. It has set up a joint venture to promote agri-biotechnology parks and a model park of such a nature has come up in Jaipur on 20 acres of land.

The campaign launched by the Foundation in the Sheldiawati region of the state in favour of organic farming has resulted in reduction of cultivation costs and improvement in the quality of the produces. Small and marginal farmers in the districts of Sikar and Jhunjhunu, who had been complaining of degradation of their agricultural land and declining productivity, were benefited. The efforts of the Foundation to popularise the use of vermi-compost in place of the chemical fertiliser by creating awareness and imparting training to the farmers have been successful.

The Foundation too has a large facility to produce vermi-compost. Application of vermi-compost reduces irrigation, increases the flavour of the products and results in a decline in the damage to the crops by insects. The Foundation is presently engaged in developing techniques to enrich vermi-compost through microorganisms to make it suitable for location and crop specific application. The promotion of organic farming is encouraged through natural resource management based on ecology protection and sustainable agricultural methods.

In Kerala, the POABS Organic Estate at Nelliyampathy is engaged in the organic cultivation of several agricultural produces. Research on organic farming methods including manures and pesticides is also undertaken there. Liquid manures, mixtures of slurry, cow urine and some herbal preparations are also made for use in the estate. Several people have taken to organic farming in the state fully convinced of its beneficial effects on man and nature.

A farmer who took organic farming as a mission in the district of Kottayam was an example. His estate grew only green crops, from vegetables to rubber. When the rubber plantation was raised on organic methods, the yields were low during the initial years. On the application of the organic inputs, they increased and after three years the yields were on par with the conventional rubber trees. An initiative to make the state of Kerala fully organic has begun with the formulation of a draft policy in 2003. A workshop on Organic Sustainability of Kerala—A Global Model was organised jointly by the state agriculture department and the Confederation of Indian Industry to deliberate on the organic farming in Kerala.

The government policy will focus on conversion of land, produces and budgetary support. Assistance will also be available for certification and inputs, promoting the local certifying bodies, development of agronomic practices, extension support and training to farmers. Creation of consumer awareness, quality considerations, and emphasis on income rather than on yield will also be part of the policy. A campaign to promote organic farming on the lines of the programme on Literate Kerala is proposed to be launched.

Organic farming, it is suggested, can be promoted among the educated youth who are presently averse to conventional agriculture. The Peerumedu Development Society (PDS) is a cooperative society in the Idukki district, which promotes organic farming for the last 12 years. Pepper, cardamom, coffee, tea, nutmeg, clove, etc. are the crops raised by about 20000 farmers of the PDS who are encouraged to adopt organic farming. PDS collects the products from its member farmers and exports too. It has also started an organic tea processing factory in district.

A view currently gaining ground in the state is that organic farming has almost become necessary for the Kerala farmer, reeling under the onslaught of the highly sensitive international markets for agriculture products. The close linking of solid waste management with bio-manure production is opening up new avenues in city planning in Kerala. The State has short listed three organisations for conducting solid waste management programmes.

The possibilities of making fuel pellets out of the rejected wastes are also being explored. 'Jaivam' or organic is the catchword now gaining popularity in Kerala. The tide seems to be turning in favour of bio-manure, bio-pest control and bio-disease control in all forms of agriculture. The state controlled Vegetable and Fruits Promotion Council, Keralam (VFPCK) is also taking slow steps in promoting the 'bio' trend as against the overriding importance given to the chemical methods of farming.

'Infam' is an association of agriculturists functioning in the state. As a part of its efforts to promote organic farming, it is planning to launch a producers' company for marketing of organic produces with a view to assist the growing trend of organic farming in the state. In the initial stage, the functioning of the company will be confined to the Wayanad district. The company will have about 3200 farmers of the district as members who have decided to switch over to organic mode of cultivation. Steps are also being taken to forge collaboration with the Indian Institute of Rural Management (IRMA), Anand.

The proposed tie up with IRMA will be for technical help and also for marketing of the products. The Infam has conducted about 300 training programmes in Wayanad district, which had motivated the farmers to adopt organic farming methods of cultivation. The MS Swaminathan Research Foundation, Chennai that has been operating in the area, assists the movement. Wayanad district is on the verge of an environmental calamity due to the excessive use of chemical fertilisers and pesticides that depleted the soil beyond redemption. The top soil is fast losing its fertility and is becoming barren in many areas. About 2100 banana farmers used tonnes of Furidan, an insecticide annually in the district alone.

The increased use of chemicals was also posing a health hazard. A survey conducted by the Infam among the school children in Thavingal Panchayath recently revealed that 250 of them were cancer patients. The cancer disease in the region is alarmingly on the rise. Karnataka has finalised the policy on organic farming and thus has become the first state in the country to adopt it. It encourages the farmers to adopt organic farming and gradually give up the use of chemical fertilisers and pesticides. The government has earmarked Rs 20 crores in the state budget of 2004-05 for the purpose.

The organic farming movement in the state has reached a stage in which the creation of markets for the products has become essential. Association for Promotion of Organic Farming (APOF) at Bangalore consists of 150 voluntary members. It has been promoting organic farming by creating awareness, inspecting and giving organic certification to the farms. Organic farm production in the state is not very significant. However, some NGOs and individuals are buying organic produces although the movement to convert the land to organic farming has not been successful.

There is no organised market for organic produces and if not sold out in the right time, they are left with the conventional produce at the same prices. At the same time, while the farmers are interested in the organic cultivation the consumers are willing to pay for such products. But the link between the two simply does not exist. Consumers demand consistent supplies of good quality organic products and the farmers want an assured market for such produces. APOP is making the farmers aware of the organic farming methods, particularly about the preparation of the soil for the purpose.

The principles behind the healthy soil through the use of the on-farm inputs and compost making are explained to them. Purushotthom Rao who owned 10 acres of land in the Shimoga district took up organic farming in 1989. He raised coconut and paddy along with several other crops in place of arecanut which was destroyed due to strong winds. The yield of coconut he obtained was 80 per cent more than that of his conventional farming friends. The cost of paddy cultivation incurred was 80 per cent lesser than his said friends.

An enterprising farmer practicing organic farming could export several products to the European market. His firm sent 5000 kg of black pepper, 1000 kg of nutmeg, 500 kg of mace, 4000 kg of white pepper, 1000 kg of clove, 300 kg of vanilla, 2000 kg of turmeric, 2000 kg of ginger, 5000 kg of white hibiscus, 80000 kg of henna and 500 kg of cardamom. His firm is certified by a European agency as organic since 1994. The state of Manipur in the north-east of India has decided to encourage organic agriculture.

The north-east region with its unique characteristics and agricultural practices can be a potential area for the introduction of organic farming. Manipur is aiming at the encouragement of the eco-friendly farm practices and pollution free industry in the state. Dependence of the agro-chemicals will be avoided and factories and industries causing harm to plants and humans will be discouraged. The emphasis on protection of environment is in perfect harmony with the identification of the entire north-east as the mega diversity spot containing rich reserves of flora and fauna. Given the geographical location and agro-climatic zones, the state gives importance to horticulture and agro-industries.

The National Horticulture Board has identified 2.77 lakh hectares for the exclusive development of horticulture in the state. It has been decided to encourage organic farming in this development area. The state has proposed to the setting up of an exclusive export zone for organic products in collaboration with the Agriculture Processed Food Export Development Authority (APEDA). Tamil Nadu plans to encourage organic farming in horticulture and plantation crops to increase the income of the farmers of the state.

The state government also wants to promote organic cultivation of fruits, vegetables and tea. The Tamil Nadu Agricultural University has established a model organic farm on a 2.5 hectare area in the campus. The National Bank for Agriculture and Rural Development (NABARD) has decided to promote organic cultivation of horticultural crops in the mango producing areas of southern Tamil Nadu districts. Theni and Dindigul are major mango growing areas and the fact that the state government has decided to promote an agri export zone for mangoes in these districts will quicken the efforts for organic farming.

The aim is to take the benefits of the premium prices for organic products in the developed countries through exports of mangoes. NABARD has funded a farmer in Theni district to the tune of Rs 3.5 crores for organic mango cultivation on a 252 acre area. Identification of land for organic cultivation was a difficult process and it took one year for the farmer to do so and another three years to convince the banks. Search for the ideal planting materials took several months and he incurred an annual cost of Rs 46000 for certification.

A seminar organised at Shri Murugappa Chettiar Research Centre (MCRC) at Chennai focused on the need for natural cultivation of food products. Application of farm manure, biological pest control, companion planting, inter cropping, introduction of beneficial microbes and fungies were advocated for natural food farming. The importance of earthworms and the harm caused to them by chemical farming were highlighted at the discussions. In the history of organic farming in Tamil Nadu, the efforts made in 1970s to revive the traditional methods of agriculture at Chennai were praiseworthy.

A study held by MCRC in the city revealed that 58 per cent of the people were aware of the benefits of organic farming and 82 per cent were ready to buy organic foods if the prices were competitive. A few farmers in Puliangudi village in Tirunelveli district have successfully adopted a package of eco-friendly technologies in the paddy cultivation. It is found that these practices show good results on the indigenous rice varieties. The cost of cultivation has substantially been reduced and the farmers receive a premium for the organic rice.

The cost of cultivation worked out comes to about Rs 8,750 per hectare. The price of rice obtained was Rs 30 per kg. The cultivation turned out to be rewarding economically besides being environmentally acceptable. Several farmers are coming forward to practice the method of organic farming after witnessing the results. Another farmer of the same village revealed that he got about 9,250 kg of paddy per ha and no plant protection was done after adoption of the new method. His cost of cultivation worked out was about Rs 12,500 per ha.

In Goa, a co-operative institution, the Adarsha Krishi Sahakari Kharedi Vikri Prakriya Sanstha (AKSKVPS), pioneered organic farming of cashew and coconut crops. The organisation has 1000 members and is led by its Chairman under whose leadership the organic cultivation is launched. The Cooperative Society has been receiving export orders from traders from abroad for supply of organic cashew nuts. The Society has helped more than 450 cashew growers to convert to organic farming by imparting training and other supports. It also has a processing facility for about 100 tonnes of cashew nuts during the season.

The price paid for a kg of organic raw nuts is Rs 35.50. The processed nuts are taken to Mumbai and Delhi by the Society for sale. The state of Madhya Pradesh has gone for organic soya cultivation. Plans are also ready to convert maize, wheat and pulse cultivation to organic on the principle of crop rotation. Organic soyabean is cultivated in the Malwa region of the state in the districts of Khargone and Dhar on an area of 8,700 hectares under contract farming.

Farmers who have not used chemical fertilisers and pesticides during the last two years on the land offered for organic farming are selected. This condition was necessary to restore the soil health to make the land suitable for organic cultivation. Certification of the farm has been done by SKAL, an international agency accredited to APEDA. On getting the certificate, the farm can use the logo 'India Organic' on soyabean and its products like soya oil and deoiled cake. The production is meant for both the national and international markets. It is said that the conventional yellow soyabean fetches Rs 12,000 per tonne in the local market, while the organic soyabean rules at Rs 34,000 per tonne in Delhi.

However, a price premium of 50 per cent is the least expected for the organic soya. This will be financially beneficial to the farmers, even though the productivity of conventional soyabean is put at 800-900 kg, which is higher than 500-600 kg of the organic soya. The genetically modified soya crops dominate the international market for soyabean and its products. However, the demand for organic soya is high ensuring a good premium for exports from the country.

CONSTRAINTS AND PROSPECTS

It is quite natural that a change in the system of agriculture in a country of more than a billion people should be a well thought out process, which requires utmost care and caution. There may be several impediments on the way. An understanding of these problems and prospects will go a long way in decision making.

Constraints

The most important constraint felt in the progress of organic farming is the inability of the government policy making level to take a firm decision to promote organic agriculture. Unless such a clear and unambiguous direction is available in terms of both financial and technical supports, from the Centre to the Panchayath levels, mere regulation making will amount to nothing. The following are found to be the major problem areas for the growth of organic farming in the country:

Lack of awareness

It is a fact that many farmers in the country have only vague ideas about organic farming and its advantages as against the conventional farming methods. Use of bio-fertilisers and bio pesticides requires awareness and willingness on the part of the farming community. Knowledge about the availability and usefulness of supplementary nutrients to enrich the soil is also vital to increase productivity. Farmers lack knowledge of compost making using the modern techniques and also its application.

The maximum they do is making a pit and fill it with small quantities of wastes. Often the pit is flooded with rainwater and result is the top of the compost remains under composted the bottom becomes like a hard cake. Proper training to the farmers will be necessary to make vermi-compost on the modern lines. Attention on the application of composts/organic manure is also lacking. The organic matter is spread during the months when the right moisture level is absent on the soil. The whole manure turns into wastes in the process. The required operation is of course labour intensive and costly, but it is necessary to obtain the desired results.

Output marketing problems

It is found that before the beginning of the cultivation of organic crops, their marketability and that too at a premium over the conventional produce has to be assured. Inability to obtain a premium price, at least during the period required to achieve the productivity levels of the conventional crop will be a setback. It was found that the farmers of organic wheat in Rajasthan got lower prices than those of the conventional wheat. The cost of marketing of both types of products was also same and the buyers of wheat were not prepared to pay higher prices to the organic variety.

Shortage of biomass

Many experts and well informed farmers are not sure

whether all the nutrients with the required quantities can be made available by the organic materials. Even if this problem can be surmounted, they are of the view that the available organic matter is not simply enough to meet the requirements. The crop residues useful to prepare vermi-compost are removed after harvest from the farms and they are used as fodder and fuel. Even if some are left out on the farms termites, etc destroy them.

Experiments have shown that the crop residues ploughed back into soil will increase productivity and a better alternative is conversion into compost. The small and marginal cultivators have difficulties in getting the organic manures compared to the chemical fertilisers, which can be bought easily, of course if they have the financial ability. But they have to either produce the organic manures by utilising the bio-mass they have or they have to be collected from the locality with a minimum effort and cost. Increasing pressure of population and the disappearance of the common lands including the wastes and government lands make the task difficult.

Inadequate supporting infrastructure

In spite of the adoption of the NPOP during 2000, the state governments are yet to formulate policies and a credible mechanism to implement them. There are only four agencies for accreditation and their expertise is limited to fruits and vegetables, tea, coffee and spices. The certifying agencies are inadequate, the recognised green markets are non-existent, the trade channels are yet to be formed and the infrastructure facilities for verification leading to certification of the farms are inadequate.

High input costs

The small and marginal farmers in India have been practicing a sort of organic farming in the form of the traditional farming system. They use local or own farm renewable resources and carry on the agricultural practices in an ecologically friendly environment. However, now the costs of the organic inputs are higher than those of industrially produced chemical fertilisers and pesticides including other inputs used in the conventional farming system. The groundnut cake, neem seed and cake, vermi-compost, silt, cow dung, other manures, etc. applied as organic manure are increasingly becoming costly making them unaffordable to the small cultivators.

Marketing problems of organic inputs

Bio-fertilisers and bio-pesticides are yet to become popular in the country. There is a lack of marketing and distribution network for them because the retailers are not interested to deal in these products, as the demand is low. The erratic supplies and the low level of awareness of the cultivators also add to the problem. Higher margins of profit for chemical fertilisers and pesticides for retailing, heavy advertisement campaigns by the manufacturers and dealers are other major problems affecting the markets for organic inputs in India.

Absence of an appropriate agriculture policy

Promotion of organic agriculture both for export and domestic consumption, the requirements of food security for millions of the poor, national self-sufficiency in food production, product and input supplies, etc. are vital issues which will have to be dealt with in an appropriate agriculture policy of India. These are serious issues the solution for which hard and consistent efforts along with a national consensus will be essential to go forward. Formulation of an appropriate agriculture policy taking care of these complexities is essential to promote organic agriculture in a big way.

Lack of financial support

The developing countries like India have to design a plethora of national and regional standards in attune with those of the developed countries. The adoption and maintenance of such a regulatory framework and its implementation will be costly. The cost of certification, a major component of which is the periodical inspections carried out by the certifying agencies, which have freedom to fix the timings, type and number of such inspections appears to be burdensome for the small and marginal farmers.

Of course, the fees charged by the international agencies working in India before the NPOP were prohibitive and that was a reason for the weak response to organic agriculture even among the large farms in the country. No financial support as being provided in advanced countries like Germany is available in India. Supports for the marketing of the organic products are also not forthcoming neither from the State nor from the Union governments. Even the financial assistance extended to the conventional farming methods are absent for the promotion of organic farming.

Low yields

In many cases the farmers experience some loss in yields on discarding synthetic inputs on conversion of their farming method from conventional to organic. Restoration of full biological activity in terms of growth of beneficial insect populations, nitrogen fixation from legumes, pest suppression and fertility problems will take some time and the reduction in the yield rates is the result in the interregnum.

It may also be possible that it will take years to make organic production possible on the farm. Small and marginal farmers cannot take the risk of low jaelds for the initial 2-3 years on the conversion to organic farming. There are no schemes to compensate them during the gestation period. The price premiums on the organic products will not be much of help, as they will disappear once significant quantities of organic farm products are made available.

Demand for organic products

The demand for organic products is high in the advanced countries of the west like USA, European Union and Japan. It is reported that the US consumers are ready to pay a premium price of 60 to 100 per cent for the organic products. The upper classes in India are also following this trend as elsewhere. The market survey done by the International Trade Centre (ITC) during 2000 indicates that the demand for organic products is growing rapidly in many of the world markets while the supply is unable to match it.

India is known in the world organic market as a tea supplier and there is a good potential to export coffee, vegetables, sugar, herbs, spices and vanilla. In spite of the several initiatives to produce and export organic produces from the country, the aggregate production for export came to only about 14000 tonnes. This also includes the production of organic spices in about 1000 ha under certification.

Some export houses like Good Value Marketing Ltd and Burmah Trading Corporation are also engaged in exporting of organic fruits, vegetables and coffee from India. The country could export almost 85 per cent of the production indicating that demand is not a constraint in the international markets for organic products.

Vested interests

Hybrid seeds are designed to respond to fertilisers and chemicals. The seed, fertiliser and pesticide industry as also the importers of these inputs to the country have a stake in the conventional farming. Their opposition to organic farming stems from these interests.

Quality standards for biomanures

The need for fixing standards and quality parameters for bio-fertilisers and biomanures has arisen with the increasing popularity of organic farming in the country. There are a very large number of brands of organic manures, claiming the high levels of natural nutrients and essential elements. But most farmers are not aware of the pitfalls of using the commercially available biomanure products.

While the concept of organic farming itself lays great stress on the manures produced on the farm and the farmers' household, many of the branded products available in the market may not be really organic. Elements of chemicals slipping into the manures through faulty production methods could make the product not certifiable as organic. The process of composting which is a major activity to be carefully done is achieved usually by one of the two methods, vermi-composting or microbe composting.

While the former is ideal for segregated waste material without foreign matter, microbe composting is suitable for large scale management of solid wastes, especially in cities and metres. Even though the farmers are using manure produced by different methods, proper parameters for biomanure are yet to be finalised. Most farmers are still unaware of the difference between biomanure and bio-fertiliser, it is point out.

While biomanure contains organic matter, which improves the soil quality, bio-fertilisers are nutritional additives separated from the organic material, which could be added to the soil, much like taking vitamin pills. Biofertilisers do nothing to enhance soil quality while the loss of soil quality has been the major problem faced by farmers these days.

Improper accounting method

An understanding of the real costs of erosion of soil and human health, the loss of welfare of both humans and other living things and the computation of these costs are necessary to evaluate the benefits of organic farming. These costs will have to be integrated to a plan for the implementation of organic agriculture. A recent study shows the inappropriateness of the cost and return accounting methods adopted to find out the economics of the organic farming. An economic evaluation of the bad effects of inorganic agriculture and their internalisation through environmental taxes is proposed for a market based approach to promote organic farming in India.

Political and social factors

Agriculture in India is subject to political interventions with the objectives of dispensing favours for electoral benefits. Subsidies and other supports from both the Central and state governments, government controlled prices of inputs like chemical fertilisers, the public sector units' dominant role in the production of fertilisers, government support/ floor prices for many agricultural products, supply of inputs like power and water either free of cost or at a subsidised rate, etc. are the tools often used to achieve political objectives. Any movement for the promotion of organic farming in India will have to counter opposition from the sections who benefit from such policies in the conventional farming system.

The political system in a democracy like India is likely to evade the formulation of policies, which affect the interests of the voting blocks unless there are more powerful counter forces demanding changes. In the absence of alternative employment opportunities and other considerations, the organised workforce particularly in the public sector fertiliser, pesticide and seed industries is also likely to oppose moves on the part of the government to promote organic farming on a large scale.

Prospects

Indian agriculture should be able not only to maintain but also must strive to increase the production of foodgrains. It appears that given the availability of organic infrastructure, minimum efforts for conversion due to the low use of chemical farming methods and the limit of the public investment, organic farming can be progressively introduced. The potential areas and crops, which fulfill the above constraints, could be explored and brought under organic agriculture.

The rainfed, tribal, north-east and hilly regions of India where the traditional farming is more or less practiced could be considered. Agriculture production in these areas is still almost on the traditional eco-friendly lines and making the farmers aware of the methods of organic farming may not be very difficult. A strategy to prevent sudden and substantial yield losses is to convert to organic production in phases to reduce the risks during the initial years.

Chemical fertiliser is applied only in 30 per cent of the cultivated area, which is irrigated, and the remaining land is under rainfed agriculture with almost no fertiliser application. Also the rainfed area under cultivation accounts for only 40 per cent of the foodgrain production of the country. The introduction of organic farming in these areas will allay the fears of a sudden sharp decline of food production which many fear may drive the nation to food imports. Thus the demand for biomass for the production of organic manures can also be controlled in a phased manner.

Moreover, the simple technologies with low input use have been developed for dry farming and they can be transferred to the farms for organic cultivation. The resulting increases in productivity and sustainability of production will increasingly contribute to the betterment of the economic condition of the dry land farming community, which is one of the poorest in the country. An estimate indicates that about 600 to 700 million tonnes of biomass is available to be converted to manure. Such conversion increases the nutrient value from 0.3-04 to 1-2 per cent.

Attempts can also be made to increase the supply of biomass by allocating a portion of the cultivated area to grow tree manure crops. These plants can be harvested to be used for making composts. Schemes can be devised to grow green manure crops in the public lands on the lines of the social forestry programmes. Several alternatives for supply of organic soil nutrients like vermi-composts and biofertilisers exist.

Technologies have been developed to produce large quantities of these nutrients. Crop specific biofertilisers for cereals, millets, pulses and oil seeds are also available. Vermi- composting and bio-fertiliser manufacturing can be undertaken to increase the supply of organic manure to meet the demand. The basic rules and regulations for accreditation and certification of organic products are in place in India. A Congenial socio-cultural environment prevails in India for the promotion of organic agriculture.

The farmers of India had been practicing eco-friendly agriculture for centuries till the advent of the 'green revolution' which was based on the conventional farming methods prevailed in the western countries. Still many small and marginal farmers, because of many reasons, have not fully adopted the conventional farming and they follow more or less the traditional environment friendly system.

They use local or own farm derived renewable resources and manage self-regulated ecological and biological processes. This has become necessary to cultivate the acceptable levels of crop, livestock and human nutrition products and above all to protect both the crops and humans from pests and diseases through the use of bio-chemicals and bio fertilisers. Such a situation is suitable for making the farming community aware of the organic farming methods to make the switch over less troublesome.

A country like India can enjoy a number of benefits from the adoption of organic farming. The price premiums for the products, conservation of the natural resources in terms of improved soil fertility and water quality, prevention of soil erosion, preservation of natural and agro-biodiversity are major benefits. Economic and social benefits like generation of rural employment, lower urban migration, improved household nutrition, local food security and reduced dependence on external inputs will be large gains in the Indian conditions.

The protection of environment and the consequent increase in the quality of human life will be other contributions of organic farming. There is a good demand for organic products in the domestic market, which is not matched by supplies. The linkages between the two do not exist which in turn discourage production. The wholesalers/traders play a major role in the distribution of organic produces as they originate from the small farms.

Large farmers have access to supermarkets and own stalls for distribution. Mumbai, Delhi, Kolkota, Chennai, Bangalore and Hyderbad are the major domestic markets for organic products. Prospects for organic farming in India can be gauged from the earlier mentioned experiences of soyabean cultivation initiated in Madhya Pradesh. High premium prices both in the national and international markets and the lack of supplies in the latter reveal the opportunities for India. Similarly the case of organic cotton of India is judged as having bright prospects.

Organic cotton/eco-cotton/green cotton is the cotton grown without inorganic fertilisers, pesticides and defoliants and duly certified by a recognised certifying organisation. India has tremendous potential to emerge as a world leader in organic cotton. Indian cotton textile industry is one of the oldest organised enterprises and is supported by supply of raw cotton cultivated in the country. Cotton based products come to about 35 per cent of the country's exports.

The present level of cotton productivity is only about 300 kg lint/ha against a world average of 600 kg. The low productivity is attributed to a large area under rainfed farming—almost 65 per cent of the cotton area. India' has the largest area under cotton in the world, but it is only the third in terms of production. The productivity is low even after 75 per cent of the cotton area is covered by the high yielding varieties. The reason for low productivity among others is the diseases/pest attacks on the crop as a result of the conversion of the traditional varieties to the high yielding ones.

India needs to increase cotton production to meet both the domestic and international demand. This has to come from an increase in productivity, which presently is very low. India is the only producer of all the four species of cotton in the world. Not only this, the country produces the widest quality range of cotton suitable for spinning 6's to 12's count yarn which can meet the demand from both the domestic and international textile industries.

During the first half of the last century, the country grew only the desi cotton and the American variety accounted for a small area only. The desi variety yields are low, but they require lesser moisture, resist pests and diseases to a considerable extent, and are suitable in dry lands. And this cotton was organically cultivated. The increasing demand for cotton pushed out the desi cotton from cultivation and was replaced by the high yielding varieties. The desi cotton is grown only on 25 per cent of the total area under cotton at present.

The conventional method of cultivation of cotton needs intensive production technologies requiring very high levels of inputs like fertilisers, insecticides, and water for irrigation. The inputs are often excessively used causing ecological problems besides diminishing rate of returns. Use of pesticides in the cotton crop is very intense and several chemicals are used indiscriminately ignoring the environment. Even the chemicals, banned/restricted for use in the advanced countries, are used in the Indian cotton crop.

Organic cultivation of cotton under certification will be profitable as there is a strong demand for eco-cotton in many Western and Asian countries. There are about 12 countries producing eco-cotton. USA, Greece, Israel, Peru, Egypt, Turkey, China and Australia are important eco-cotton producing countries. Eco-cotton commands a price higher than 30 to 40 per cent of the conventional cotton. However, the cotton growers in the country will take up organic cultivation only if the yield loss if any as a result of the switch over is compensated by high prices and the technology is feasible and financially viable.

Organic cotton can be produced in India easily in those areas where cotton is raised under the dry farming conditions. These areas are under the desi variety and as such can be converted into organic cotton farming easily. But the potential areas are spread over several states of western and southern India. Coloured cotton was cultivated in Andhra Pradesh prior to 1950. High quality hand woven fabrics were made from desi cotton once upon the time. But white cotton has replaced the coloured ones during the last fifty years. Now there is a demand world over for organically produced coloured cotton and India is in a position to cater to the international market.

REFERENCES

Alam, Anwar and Wani, Shaftq, A., "Emerging Need for Organic Agriculture and Strategies for its Optimization," in *Proceedings of National Seminar on Organic Products arid Their Future Prospects*, Sher-e-Kashmir University of Agricultural Sciences and Technology, Srinagar, 2000, pp 95-103.

Geier, Bemwad, "International Federation of Organic Agriculture Movements," in *Sustainable Agriculture Solutions: The Action report of the Sustainable Agriculture Initiative*, Novello Press, London, 1999.

Rao, PS, "Marketing of Organic Produce of Wheat in Rajasamand District of Rajasthan," *Indian Journal of Agriculture Marketing*, Conference Number Special, 2003.

Singh, GR, "Organic Farming for Sustainable Agriculture," *Indian Farming*, June, 2001, pp. 12-14.

Veeresh, GK, "Organic Farming Ecologically Sound and Economically Sustainable," *Plant Horti Tech*, 1(3), Nov-Dec, 1999.

6

Problems and Constraints in Organic Farming

According to the Food and Agriculture Organisation (FAO), sustainable agriculture "is the successful management of resources for agriculture to satisfy changing human needs while maintaining or enhancing the quality of environment and conserving natural resources". All definitions of sustainable agriculture lay great emphasis on maintaining an agriculture growth rate, which can meet the demand for food of all living things without draining the basic resources.

Organic farming is one of the several approaches found to meet the objectives of sustainable agriculture. Many techniques used in organic farming like intercropping, mulching and integration of crops and livestock are not alien to various agriculture systems including the traditional agriculture practiced in old countries like India. However, organic farming is based on various laws and certification programmes, which prohibit the use of almost all synthetic inputs, and health of the soil is recognised as the central theme of the method.

Adverse effects of modern agricultural practices not only on the farm but also on the health of all living things and thus on the environment have been well documented all over the world. Application of technology, particularly in terms of the use of chemical fertilisers and pesticides all around us has persuaded people to think aloud. Their negative effects on the environment are manifested through soil erosion, water shortages, salination, soil contamination, genetic erosion, etc.

Organic farming is one of the widely used methods, which is thought of as the best alternative to avoid the ill effects of chemical farming. There are several definitions of organic farming and the one given by the US Department of Agriculture (USDA) is considered the most coherent and stringent. It is defined as 'a system that is designed and maintained to produce agricultural products by the use of methods and substances that maintain the integrity of organic agricultural products until they reach the consumer.

This is accomplished by using substances, to fulfill any specific fluctuation within the system so as to maintain long term soil biological activity, ensure effective peak management, recycle wastes to return nutrients to the land, provide attentive care for farm animals and handle the agricultural products without the use of extraneous synthetic additives or processing in accordance with the act and the regulations in this part'. The origin of organic farming goes back, in its recent history, to 1940s.

During this period, the path breaking literature on the subject published by J.I. Rodale in the United States, Lady Balfour in England and Sir Albert Howard in India contributed to the cause of organic farming. The farming being practiced for the last three decades in India has increasingly been found non-sustainable. The system is oriented towards high production without much concern for ecology and the very existence of man himself.

There are three categories of opinions about the relevance of organic farming for India. The first one simply dismisses it as a fad or craze. The second category, which includes many farmers and scientists, opines that there are merits in the organic farming but we should proceed cautiously considering the national needs and conditions in which Indian agriculture functions. They are fully aware of the environmental problems created by the conventional farming. But many of them believe that yields are lower in organic cultivation during the initial period and also the cost of labour tends to increase therein. The third one is all for organic farming and advocates its adoption wholeheartedly. They think that tomorrow's ecology is more important than today's conventional farm benefits.

However, among many a major reservation, the profitability of organic farming vis a vis conventional farming, is the crucial one from the point of view of the Indian farmers, particularly the small and marginal. Organic farming involves management of the agro-eco system as autonomous, based on the capacity of the soil in the given local climatic conditions. In spite of the ridicule poured out on organic farming by many, it has come to stay and is spreading steadily but slowly all over the world. India has been very slow to adopt it but it has made Inroads into our conventional farming system.

One advantage we have here is the fact that the farming techniques practiced in this country before the advent of the green revolution were basically eco-lriendly and they have not faded away from the memories of the present elder generation of our farming community. India's options in finding out an alternative method to the conventional farming are limited. The World Trade Organisation (WTO) deadline for banning exports of vegetables and horticulture products based on inorganic farming will expire in 2005. Sensing the importance, the Central and state governments have taken several initiatives to popularise organic farming in the country.

Howard's Agricultural Testament draws attention to the destruction of soil and deals with the consequences of it. It suggests methods to restore and maintain the soil fertility. The study contains a detailed deposition of the famous Indore method of maintaining soil health. The reasons and sources of the erosion of soil fertility and its effect on living things are discussed. The criticism of the agriculture research and examples of how it had to be carried out to protect soil and its productivity are discussed in detail.

Bemwad Geier is of the opinion that there is no other farming method so clearly regulated by standards and rules as organic agriculture. The organic movement has decades of experience through practicing ecologically sound agriculture and also in establishing inspection and certification schemes to give the consumers the guarantee and confidence in actuality. Organic farming reduces external inputs and it is based on a holistic approach to farming. He describes the worldwide success stories of organic farming based on the performance of important countries in the west.

The magnitude of world trade in organic farming products is also mentioned. To the question of whether the organic farming can feed the world, he says that neither chemical nor organic farming systems can do it; but the farmers can. Save and Sanghavi are of the view that after their intensive experiments with organic farming and narrating the results to the informed, it is time that the governments and farmers are brought around. They firmly state that the economic profitability of organic farming can be proved.

Four crops of banana grown by the natural way on the same farm by them are compared with those produced by the conventional way. While the natural farm yielded 18 kg of banana in the first round, the conventional one gave 25 kg. 30 kg was the yield at the second round on both the farms. However, on the third round, the natural farm gave 25 kg, the conventional one yielded only 20 kg. The results on the fourth round were stunning—the plants on the conventional farm died out; but the natural ones gave 15 kg on an average. Thus, the aggregate output was 88 kg on the natural farm and 75 kg on the conventional one.

While, the natural banana commanded a price of Rs 2.50 per kg, the conventional one could fetch only Rs 1.75 per kg. This has been the major reason for the substantial net profit (Rs 154) earned from the cultivation of natural banana. The expenses incurred were Rs 66 and Rs 105 for the natural and conventional bananas respectively. However, a stringent cost and return analysis representing a larger sample size will be necessary to draw meaningful conclusions. It should be born in mind that the output obtained from the natural banana farm was also because of the accessibility to the inputs and expertise, which the authors happened to possess.

Farmers placed in less advantageous positions may not derive such results. The price advantage to the natural organic farming products will also taper off when the supply increases. The environmental costs and returns have to be internalised and it is quite possible that the organic farming will prove to be a far better alternative to the conventional one. However, these aspects will have to be built into a scientific and tight economic reasoning, among others. Kaushik analyses the issues and policy implications in the adoption of sustainable agriculture.

The concept of trades off has a forceful role to play in organic farming both at the individual and national decision making levels. Public vis-a-vis private benefits, current vis-a-vis future incomes, current consumption and future growths, etc. are very pertinent issues to be determined. The author also lists a host of other issues. While this study makes a contribution at the conceptual level, it has not attempted to answer the practical questions in the minds of the farmers and other sections of the people. Sharma makes a case for organic farming as the most widely recognised alternative farming system to the conventional one.

Veeresh opines that both high technology and sustainable environment cannot go together. Organic farming is conceived as one of the alternatives to conventional agriculture in order to sustain production without seriously harming the environment and ecology. However, he says that in different countries organic farming is perceived differently. While in the advanced countries, its focus is on prevention of chemical contamination, we, in countries like India are concerned of the low soil productivity.

Even the capacity to absorb fertilisers depends on the organic content of the soil. The principles of organic farming are more scientific than those of the conventional. India's productivity of many crops is the lowest in the world in spite of the increase in the conventional input use. The decline in soil nutrients, particularly in areas where the chemical inputs are increasingly being used in the absence of adequate organic matter is cited as a reason for low productivity.

Doubts about the availability of massive sources of organic inputs also exist. He advocates an advance to organic farming at a reasonable pace and recommends conversion of only 70 per cent of the total cultivable area where unirrigated farming is in vogue. This 70 per cent supplies only 40 per cent of our food production. While this analysis has several merits, it is more addressed to the policy makers and less to the farmers.

Sankaram Ayala is of the view that almost all benefits of high yielding varieties based farming accrue mostly in the short term and in the long term they cause adverse effects. There is an urgent need for a corrective action. The author rules out organic farming based on the absolute exclusion of fertilisers and chemicals, not only for the present, but also in the foreseeable future. There ought to be an appropriate blend of conventional farming system and its alternatives. The average yields under organic and conventional practices are almost the same and the declining yield rate over time is slightly lower in organic farming.

Decreased aggregate outputs, increased farm income and increased consumer prices are other results the model gives. While the details about this US analysis are not known, its relevance to India where we already have the lowest yields of a number of crops under the conventionai system appears to be open. Singh and others, recording the experiments on rice-chick pea cropping sequence using organic manure, found the yields substantially higher compared to the control group. Similar results were obtained for rice, ginger, sunflower, soyabean and sesame.

Ahn Jongsung opines that organic agriculture is economically viable. The author gives emphasis on marketing the organic products on the basis of reputation and credibility. In Japan, the farmers sell the produces directly to the consumers. The Kenyan farmers have seen that in organic farming, costs go down and profits increase. A farmer from UP who allotted a portion of his land exclusively for organic farming found that the yields of sugarcane, rice, wheat and vegetables were lower than those under chemical farming.

An Englishman, settled in Tamil Nadu, who runs an organic farm in 70 acres planted with coffee, citrus, other fruits, rice, pepper and vegetables says that he does not earn a profit and does not have confidence in organic farming. Somani and others have published a collection of 42 papers presented at a National Seminar on Natural Farming. Korah Mathen recounts several problems in evolving representative and rigorous yardsticks for comparison between modern and alternative farming. Yields cannot be compared, because of monoculture nature of chemical farming with those of multi crops raised under organic/ natural farming.

Economic analysis is also problematic because one has to quantify the intangibles. He advocated the resource use efficiency analysis. But the question of profitability of different systems of farming seems difficult to be examined in the absence of an economic analysis although the author does not rely upon it. Save found that after three years of switching over to natural cultivation, the soil was still recovering from the after effects of chemical farming.

When the soil regained its health, production increased and the use of inputs decreased. The farm, which was yielding 200 to 250 coconuts per tree, gave 350 to 400 per annum. Rahudkar and Phate narrate the experiences of organic farming in Maharashtra. Individual farmers growing sugarcane and grapes, after using vermi compost, saw the soil fertility increased, irrigation decreased by 45 per cent and sugarcane quality improved. The authors say that net profits from both the sugarcane and grape crops are high in organic farms.

The foregoing overview of the literature makes it clear that opinions about organic farming are divided both among the farmers and experts. Disputes about the profitability and yield increases in organic farming are acute, but there Is a consensus on its eco-friendly nature and inherent ability to protect human health. There are strong views for and against organic farming. Those who are totally against it are prepared to ignore the ill effects of the conventional farming system.

There are many who while approving organic agriculture, want a mixture of both the systems or advocate a careful approach by proceeding slowly towards the conversion of the conventional farms into organic. The questions about the yield and financial viability are crucial from the point of view of farmers; but they remain unanswered to a large extent. The study of a geographical area in the country endowed with a large number of resources for organic farming, but has not made any significant stride towards it, is also not found in the literature overview.

RELEVANCE OF ORGANIC FARMING

The relevance and need for an eco-friendly alternative farming system arose from the ill effects of the chemical farming practices adopted worldwide during the second half of the last century. The methods of farming evolved and adopted by our forefathers for centuries were less injurious to the environment.

People began to think of various alternative farming systems based on the protection of environment which in turn would increase the welfare of the humankind by various ways like clean and healthy foods, an ecology which is condusive to the survival of all the living and non-living things, low use of the non-renewable energy sources, etc. Many systems of farming came out of the efforts of many experts and laymen. However, organic farming is considered to be the best among all of them because of its scientific approach and wider acceptance all over the world.

International Scene

The negative effects of modem chemical based farming system were first experienced by those countries, which introduced it initially. So, naturally, it was in those countries organic farming was adopted in relatively large scales. There are very large organisations promoting the organic farming movement in European countries, America and Australia etc. These organisations, for example, the International Federation of Organic Agriculture Movements (IFOAM) and Greenpeace have studied the problems of the chemical farming methods and compared the benefits accruing to the organic farming with the former. Organic farming movements have since spread to Asia and Africa too.

IFOAM was founded in France in 1972. It spearheads and coordinates organic farming efforts the world over by promoting organic agriculture as an environment friendly and sustaining method. It focuses on organic farming by highlighting the minimum pollution and low use of non-renewable natural resources through this method. It has about 600 organisational members spread over about 120 countries including India.

IFOAM undertakes a wide range of activities related to organic farming such as exchanging knowledge and thoughts among its members; representation of the movement in governmental, administrative and policy making forums in the national and international arena; updating of production, processing and trading standards; formulation and coordination of research projects; and holding of international conferences and seminars. IFOAM participates in the activities related to organic farming under the auspices of the United Nations and keeps active contacts with several international NGOs.

The Food and Agriculture Organisation (FAO) of the United Nations provides support to organic farming in the member countries. It also attempts the harmonisation of national organic standards, which is absolutely essential to increase international trade in organic products. The FAO has, in association with the World Health Organisation (WHO), evolved the Codex Alimentarius for organic products.

Organic farming has several advantages over the conventional one apart from the protection of both the environment and human health. Improved soil fertility, better water quality, prevention of soil erosion, generation of rural employment, etc. are some of them.

The concept of quality food has undergone a drastic change over the past few decades. It does give emphasis on the characteristics of the end product, but the process and method of production and transport are now considered equally important. Not only the importers but also the domestic retailers have their own quality specifications or standards, which in many cases are tougher than those of the government regulations. Consumers have become health conscious and are willing to pay for the clean, healthy and natural food.

Many developed countries have various support programmes to help organic farming with financial incentives and technical guidance. The organic food market in the world has grown rapidly in the past decade. International trade in organic foods showed an annual growth rate of about 20-22 per cent during this period. Many retail chains and supermarkets in advanced countries are accorded with 'green status' to sell organic foods.

The organic food processing industry is considered nature friendly and thus encouraged. The US food market tops with a value of $8.0 billion, followed by Germany and Japan with $2.5 billion each while France accounts for $1.25 billion. Moreover, the organic foods command higher prices which often range from 10 to 100 per cent more than the conventional ones.

The important organic products traded in the international market are dried fruits and nuts, processed fruits and vegetables, cocoa, spices, herbs, oil crops and derived products, sweeteners, dried leguminous products, meat, dairy products, alcoholic beverages, processed foods and fruit preparations. Cotton, cut flowers, animals and pot plants are major non-food products in the world markets.

Organic farming is bound to grow around the world as many countries are developing their own standards and regulations. The US and the European Union have very comprehensive National Organic Programmes and the early nineties have seen organic farming regulations in Japan, Canada and Australia. New Zealand, Israel and Brazil have adopted the organic standards equivalent to those of USA and the European Union. China, Thailand,

South Korea, Philippines, Turkey and Mexico have established certifying agencies. India too has adopted the National Programme for Organic Production (NPOP) with national standards.

There were about 35,000 certified organic farms in 17 countries during 1998 covering an area of more than a million hectares. A survey held recently revealed that 6-7 per cent farms in USA and Germany are certified organic farms. Many organic inspection and certifying agencies like Crop Improvement Association (OCIA), IFOAM, Ecocert, BCS, SKAL. KRAV, IMO, FVO, QAI, Oregon, Tilth and NASAA have strong international presence and have even developed the ability to assist marketing of organic products world wide. Certification of products is the foundation on which organic agriculture rests. Its role in production and marketing of organic foods becomes a significant corner stone in the creation of a confident consumer community.

The need for a trustworthy and healthy organic market is important as the consumers are becoming increasingly separated from the food production process. The void is filled in by a plethora of processing, distribution and marketing intermediaries. The consumers should possess knowledge and confidence that the products they buy are truly organic. Such environment can be created by an organised system of inspection and certification supported by law and regulations of the state.

Organic products enjoy trust based on the label or logo approved by the certification process. The aim of organic certification is, therefore, to (a) provide an identifiable label (a logo or a symbol) and (b) give an assurance to the consumers that the product is truly organic.

It is widely recognised the world over that the certification of organic products should be based on the following principles:

— Organic production and processing standards should be clearly laid down.
— The conformation of production and processing to these stands must be verified.
— Organic labels should be permitted only to those produces, which are found conforming to the set standards.

Thus, a label on an organic product conveys that the manufacturer has a license for organic production, an independent agency has inspected the production/processing practices followed by the producer, and compliance of the set standards is made. Such an assurance becomes crucial in the generation of consumer confidence in the organic products, particularly as they are costlier than the conventional ones. The marketing agents of organic products also need such an assurance to maintain fair trade practices essential to the growth and sustainability of the demand for such products.

Organic farming

Organic farming has spread to about 100 countries around the world. An estimate in 2004 puts about an area of 24 million hectares under oi'ganic farming worldwide. Australia with its 10.5 million hectares leads the countries. However, much of this area is pastoral land for grazing. Argentina with 3.19 million hectares, Italy with 1.83 million hectares and USA with 0.95 million hectares follow suit.

The importance of organic farming is growing in many countries. Austria and Switzerland have about 10 per cent of their food system under organic agriculture. The annual growth of organic farming is estimated to be about 20 per cent in USA, France, Japan and Singapore.

DEVELOPED COUNTRIES

Great strides have been made in organic farming by the

western developed countries. We very often tend to put small farms in harmony with the organic farming system imagining that the large farms are not attuned for its adoption. In USA, there are large organic farms coming to thousands of acres and they raise crops using organic methods completely avoiding the use chemicals for manuring and controlling of pests and weeds. They implement crop rotations and lay emphasis on the timing of cultural operations.

USA

Organic farming did spread fast in USA during the 1990s. The growth was in response to the rising demand for organic products both in the national and international markets. The US agriculture farms view organic farming as a method to reduce input costs and use of non-renewable resources, which increases accessibility to the high value markets and assures premium prices for the produces and above all increased farm income.

The US organic farming system uses eco-friendly agriculture practices in cultural and pest management and avoids the application of chemicals. Antibiotics and hormones are not used in livestock production. The growth of consumer demand for organic products has also resulted in the specialisation in growing, processing and marketing of these products. The implementation of national organic standards in 2002 by the USDA has further boosted the demand for organic products in the country.

Because of increasing prices and ineffectiveness of chemicals, many farms in USA do not favour intensive chemical based agriculture. A survey of certified organic farms revealed that more than 60 per cent of them grew vegetables and they had livestock for manure to use in the farm. The farmers had identified three problem areas, which required study. They were (1) the need for increasing demand for organic products, (2) the influence of cultural practices on quality and nutritional value of the products and (3) the relationship between plant nutrition and resistance to pests.

In another study of consumer demand for organic foods, it was found that about 30 per cent of shoppers used some organic food in a year, while 10 per cent used it once a week. The value of sales of organic products in US was $1.5 billion in 1992 and $8 billion in 2000. Many methods to promote sales and consumption of organic food have been introduced. The US Senate approved a bill to make a part of the Federal School Lunch Programme organic. Organic food shows are held annually to give awards for the best organic foods and drinks. The Federal government encourages the planting of native plants for Municipal planting programmes.

The government has initiated polices to reduce the application of pesticides and labelling food to indicate the amount of pesticides used to produce the commodity is compulsory. There are about 17 organic certification agencies in US. The certification and labelling laws are very rigid in United States. Imports of organic products must conform to the standards laid down by the USDA and naturally, the ability of many developing countries to comply with these requirements is limited.

A grower or a dealer in need of certification has to lodge a plan to an accredited agency of the USDA. There are both private and state certification programmes. The plan must contain all details in relation to the farming or handling methods and the materials those will have to be used therein. The anticipated improvements and future aims should also be a part of the plan. The farmers or dealers of wild crops should also submit a plan mentioning that the harvesting techniques to be employed will not cause any damage both to the produce and the environment.

All records related to the management practices and the materials utilised in organic production should be maintained for five years. Certification will be approved only if the crops are grown on the land free of prohibited substances for three years prior to harvest. The period of conversion is three years between the conventional and organic farming and the crops grown during this period cannot be labelled organic or any other label indicating that they are under conversion.

The Organic Foods Production Act (OFPA) contains a wide and detailed coverage of organic farming methods and materials along with the soil fertility, application of manure, crop rotation and composting. The National Organic Standards Board's (NOSB) recommended ingredients include crop residues, crop wastes from food processing operations, animal manure, yard wastes from private and municipal sources, or other vegetable by-products.

Municipal solid waste compost and sewage sludge composts and the use of prohibited material as composts/ ingredients are banned. There is a National List of acceptable and banned materials which includes pest control treatments and other agricultural inputs like fertilisers and seed treatments. An evaluation of all agricultural inputs has been made for their long term effect on the environment irrespective of the fact that they are synthetic or natural. The exporters of other countries to the US market will have to face these very tough specifications, particularly if their national standards are not equivalent to those of the USDA.

Italy

The European Union (EU) of 25 countries had 5.5 million hectares under organic farming at the end of 2002 managed by 1,60,000 organic farms. This constitutes about 3.4 per cent of the agricultural area and 1.7 per cent of farms in EU. It is estimated that the growth in the area was 9 per cent over the last year. The increases came mainly from France,

Spain and the United Kingdom. The European Commission has formulated the 'European Action Plan for Organic Food and Farming' on 10 June 2004. It proposes 21 specific actions to promote organic farming in the EU countries and the Action Plan is with the European Parliament for consideration. As said earlier, the progress achieved by Italy, Austria and Germany was impressive and as such it is discussed here.

The pioneers of organic farming in Italy had contributed immensely to the spread of organic agriculture, which is acclaimed to be one of the best in the world. Farmers, medical doctors, food processors, traders and consumers were in the forefront working hard from the 1950s for the cause of organic farming in the country. Growth of organic farming in Italy has been spectacular during the period of 1990-2000. The organically cultivated area increased from 13,000 ha to more than a million ha which came to about 7.2 per cent of the total cultivated area.

The number of farmers was about 13,00 in 1990, which shot up to 47,357 at the end of 2000. At the end of 2002, the area increased to 1,83,000 ha and the number of farms to 56,440. The reasons for such a substantial increase were the food scandals like the mad cow disease leading to the search for safe foods, the constant endeavour for alternative farming methods which were environmentally, technically and economically attractive to the farmers and the liberal subsidies.

The farmers who opted for organic methods prior to 1990 were people of non-agriculture background, people searching for novel life styles, etc. A significant proportion of them consisted of non-Italians like, Germans, Swiss, Austrians and Americans. They had the unity of purpose in establishing cultural associations in cooperation with like minded people by setting up marketing links with organic farms. However, the decade of 1990-2000 saw that the new entrants into organic farming were Italians and many of them did not have experience of the new methods. But they possessed a good knowledge of the ecology and understood the need for a marketing system.

The period also witnessed a change in the geographical distribution of area under organic cultivation. Sardinia became an important region in organic farming with the extension of subsidies to pasture lands used for sheep rearing and milk production. Sardina and Sicily account for about 44 per cent organic area in Italy. At present 41 per cent of the organic area is under pastures. Cereal crops are raised on 18.2 per cent of the area and the olive cultivation alone has the highest single crop share of 10 per cent.

The high growth of organic production led to the growth of the processing units to the extent of 142 per cent during 1997-2000 alone. Presently, there are about 42,000 processing units of which 1,300 are established by the farmers themselves. Organic farming has resulted in the setting up, conversion or expansion of new economic activities in the rural areas of Italy. The first organic shop had come up in Milan, the industrial and financial capital of Italy, during the 1970s under a consumers' cooperative venture.

The most important fact of the market distribution of organic products in Italy is the predominant position of the farmers' or consumers' cooperatives. The distribution was mostly done during the immediate post 1970s by small grocery stores and the major share was that of the direct sales in the open market places and at the farms. After 1990, specialised markets, supermarkets, franchisees, school canteens, etc. became the major channels. Supermarkets are entering the field in large numbers since 1999 with their own labels to convince the consumer of the reliability of the products in the midst of numerous labels.

Diary products have a share of 26 per cent of the consumer expenditure on organic products; fruits and vegetables accounts for 13 per cent; breads and biscuits had a share of 12 percent. A growing number of kindergartens, school and university canteens, restaurants and factory Ccinteens are opting for organic foods in the country. A survey of the motives of the consumers of organic foods revealed that every one was going for organic products to protect the health of his/her family.

The support for organic farming by the European Union by various legislations providing subsidies was instrumental in the growth of organic farming in Italy. The devolution of power to the Regional Government on agriculture policy in 1972 has resulted in different levels of subsidies in different regions/crops. There are nine organic certification bodies in Italy. Some of them were earlier cultural associations promoting organic farming. A single agency named ICEA has a share of 24 per cent of the certified farms coming to about 29 per cent of the area. These nine agencies have about 900 local offices, 1000 staff and technicians for inspecting farms, processing units and storages for giving certifications.

Universities and Public Research Centres are becoming increasingly interested in research and extension activities in relation to organic agriculture. A National School of Organic Farming has been established by the Ministry of Agriculture to impart education.

Austria

Austria is a leading country in Europe in the adoption of organic farming. At the end of 2003, about 11.6 per cent of the cultivated area which came to 2,96,154 ha was under organic agriculture. The number of organic farms came to 18,576, which was 9.2 per cent of the total number of agriculture farms in Austria. The conversion of cultivated land to organic method of agriculture was slow till 1990. The federal government introduced subsidies for organic conversion of cultivation in 1991. The government programme of agri-environment of 1995 and the conversion of grass lands into organic farms gave a great fillip to the organic movement.

There was a decrease in the conversions in 2000, but it increased substantially from 2001 to 2003. The decline was due to the difficulties in marketing the organic produces in the country. This was an indication that the emphasis on farm production was not enough to promote organic agriculture. The increasing phase from 2001 to 2003 saw about 700 farms having an area of 30,000 ha being converted. This came from lands cultivating cereals and vegetables.

One of the reasons for the change is attributed to the realisation of the benefits of organic farming by the young farmers with a better concern for ecology and understanding of economics than their parents. The organic farmers belonged to a number of associations and the latter had different approaches/objectives, regional differences and marketing techniques. About 64.5 per cent of the organic farmers were members of these associations, which functioned under two umbrella organisations, named ARGE and OelG. These associations have their own production and quality standards, which in many cases are more stringent, than those of the State.

Presently there are 11 major organic farming associations in Austria and while 6 of them come under ARGE, the remaining five and 4 marketing associations are members of OelG, the latter being the one giving importance of marketing too in the promotion of organic agriculture in Austria. The concentration of organic farms is in the regions with large areas of grass land. It is easier for the conversion of the grass land to organic farming as they used to be cultivated extensively.

Grass land farms constitute about 66 per cent of the total organic farms. The average size of an organic farm in

Austria is 16 ha. Certification of organic farms in Austria is done with the provincial government and a contract between the farm and an inspection agency is essential for registration. An approved organic label is available to guarantee the quality of the organic products. The conditions to be fulfilled by an inspection agency for accreditation are very tough in Austria. 17,500 registered organic farms receive government subsidies and about 50 per cent of it is born by the European Union and balance by the federal and provincial governments.

The market value of Austrian organic products was 2.18 billion Euro in 2000 which was 3 per cent of the total food market. About 70 per cent of these products is distributed by the conventional food chains. Now, the distribution of organic foods through the supermarket chains is gaining momentum. It is reported that the supermarkets selling organic products enjoy an improved image among their customers.

Organic farms' cooperatives also market organic products. Cooperative ventures between the organic farmers and hospitals, nursing homes, etc. have come up to sell organic products. Home delivery services and organic shops are also established, particularly in large cities. Agricultural Schools and Colleges offer training programmes/ courses in organic farming and promote scientific research. The Institute for Organic Farming at the University of Vienna is a major centre for research in organic agriculture. Development of a strategy for co-operation among the organic farmers, their associations, processors and traders is considered a major challenge in the promotion of organic farming in Austria.

Germany

Germany, where the IFOAM is located at Tholey-Theley since 1987, is a leading nation in the adoption of organic farming in Europe. The IFOAM has its Regional Group founded in 1991 with the German speaking nations (Austria, Germany, Luxemburg and Switzerland) as members. Germany has about 3 per cent of its total agricultural area under organic farming against 2 per cent of the European countries as a whole.

The country has also a large number of processors and importers of organic produces. So, it would be useful to examine the organic farming movement in that country in some detail. The history and development of organic agriculture in Germany begins from the days of bio-dynamic farming initiated in 1924. Biodynamic theory of farming treats farm as a living being which responds to non-material influences, called dynamic forces.

The proponents of biodynamic farming believe that these forces can be enhanced by biodynamic preparations. Even certification for biodynamic agriculture started in 1954 in the country. Organic-bio-dynamic farming was developed in Switzerland in the early twentieth century. Hans Mueller and his wife Maria Mueller were the spirits behind this movement in Switzerland.

Mueller had developed the method during the 1950s and the theoretical basis was provided by Hans-Peter Rusch, the German microbiologist and medical doctor. The damage inflicted to the environment by the modern farming system was clear by the end of 1960s. The ill effects of pollution were obvious and a producer organisation, named Bioland was established in 1971 in Germany. Another organisation for the organic cultivation of fruits and vegetables had been founded in 1961.

A federation for organic farming associations was established in 1988 consisting of six producer organisations. A common basic standard for organic agriculture was laid down in 1984 as a framework for individual organisation. Organic farming spread very fast in Germany during the subsequent period. State funding for organic agriculture was available from 1989. The unification of both the

Germanys in 1990 saw fast growth of organic farming in the former East German areas.

The trade sector felt the absence of a German logo for organic products and the seal 'Biosiegel' was launched by the federal government. The German government has put the goal of bringing 20 per cent of the agricultural land under organic farming by 2010 from the 3 per cent of 2002. A series of measures were introduced to achieve the goal. These steps are under the EU programme of support to organic agriculture and the federal programme for organic farming.

There are eight organic producer organisations in Germany and they represent about 60 per cent of the organic farming community. All these associations have the common organic seals for labelling the certified products, both primary and processed.

The organic producers, processors and traders of the country have formed a common organisation aiming at the propagation of achievements of organic production, protection of environment etc. There are several other organisations functioning in relation to organic farming in Germany. SOL (Foundation for Ecology and Agriculture), FiBL (Berlin), Schweisl\irh Foundation, Foundation Future Agriculture and Society for Ecological Animal Husbandry are some of the institutions contributing to the spread of organic farming in Germany.

Organic farming has gained acceptance in the former East Germany compared to the West. The reasons are the substantial changes in the German agriculture after the reunification, existence of large conservation areas in which organic farming can well be fitted into. Large areas in the eastern part are designated as disadvantaged and incentives are provided there to go for organic farming.

The arguments for and against organic farming between the organic and conventional farmers are intense in the western parts of Germany and this has to some extent slowed down the spread of the organic agriculture there. However, government regulations and the support of European Union Regulations have made the organic agriculture accepted in the country. The standards and certification process for organic farming are stricter in Germany than in other European Union countries.

The German regulations restrict certification to the whole farm and the processing standards are tough. There are private and state standards and the inspections are done according to both of them. The use of genetically engineered organisms and techniques are banned in organic agriculture. There are more than 20 inspection bodies and the 22 states of the country control them.

The German organic food market was worth about 3 billion Euro in 2000. Both the federal and state governments extend financial support to organic farming. Farmers are encouraged to adopt organic methods from 1989 and those who adopt production methods compatible with the protection of environment are given assistance. In addition to the direct subsidies, marketing assistance is also provided to the organic farms. Subsidies for producers' marketing organisations, processing and market promotion are available.

Advisory services are provided to the farmers for organic farming. Advisory services were organised by the farmers themselves during the initial period of the introduction of organic farming in Germany. Now producer associations with governmental financial support offer such services. Farmers themselves hire advisors with state support and official advisors are also available. Research and training facilities are existing in almost all the German agricultural universities and technical colleges in organic farming. Degree courses are offered in organic agriculture and organic farming research is also funded by the state. A state research institute for organic farming was established in 2000.

DEVELOPING COUNTRIES

Large sections of farmers in the developing countries are poor and their land holdings in many countries are small. Their access to external inputs is limited and their ability to improve production is low given the economic and social disabilities to which they are exposed. Globalisation and the opening up of the trade barriers among the nations have resulted in the decline of agricultural prices in the local markets to the detriment of the interests of these small farmers.

These conditions can be made a bit favourable for the introduction of organic farming in which the traditional system is found conducive for such adaptations with the state support for conversion to organic agriculture. Most of the countries lagging behind in the adoption of organic agriculture were from Asia and Africa. In terms of both in absolute and percentage area under organic farming, Tanzania, Zambia, Ghana, India and South Africa are far behind many nations in organic agriculture.

Organic farming in India

The need for organic farming in India arises from the unsustainability of agriculture production and the damage caused to ecology through the conventional farming practices. The present system of agriculture which we call 'conventional' and practiced the world over evolved in the western nations as a product of their socio-economic environment which promoted an over riding quest for accumulation of wealth. This method of farming adopted by other countries is inherently self destructive and unsustainable.

The modern farming is highly perfected by the Americans who dispossessed the natives of their farms right from the early period of the new settlers in US. The large farms appropriated by the immigrants required machines to do the large scale cultural operations. These machines needed large amount of fossil fuels besides forcing the farmers to raise the same crops again and again, in order to utilise these machines to their optimum capacities. The result was the reduction of bio-diversity and labour.

The high cost of the machines necessitated high profits, which in turn put pressure to raise productivity. Then, only those crops with high productivity were cultivated which needed increased quantities of fertilisers and pesticides. Increasing use of pesticides resulted in the damage to environment and increased resistance of insects to them. Pesticides harmed useful organisms in the soil. The monoculture of high yielding seeds required external inputs of chemical fertilisers. The fertilisers also destroy soil organisms. They damage the rhizobia that fix nitrogen and other micro organisms that make phosphates available to plants.

The long term effect was reduction of crop yields. The damaged soil was easily eroded by wind and water. The eroding soil needed use of continuously increasing quantities of fertilisers, much of which was washed/leached into surface and underground water sources. The theme of consumer welfare has become central in the economic activities in the developed countries in the world. Sustainable agriculture based on technologies that combine increased production with improved environmental protection has been accepted as absolutely essential for the maximisation of the consumer welfare.

The consumers are increasingly concerned about the quality of the products they consume and food safety has become a crucial requirement. Safety, quality and hygienic standards are increasingly being made strict. The mad cow disease and the question of genetically modified food production are the recent instances, which made the countries to tighten the laws.

Mycotoxln contamination, unacceptable levels of pesticide residues and environment degradation are the problems on which the attention is centred. Keeping the interests of the consumers, the European Union has taken tough measures including criminal prosecution to ensure food safety. Another area to increase the consumer welfare is promotion of the eco-friendly methods in agriculture. No-till, or conservation agriculture, lower input approaches of integrated pest or nutrient management and organic farming are some of them.

The Indian agriculture switched over to the conventional system of production on the advent of the green revolution in the 1970s. The change was in the national interest which suffered set backs because of the country's over dependence on the foreign food sources. The national determination was so intense that all the attention was focused on the increase in agriculture production. The agriculture and allied sectors in India provide employment to 65 per cent of the workers and accounts for 30 per cent of the national income.

The growth of population and the increase in income will lead to a rise in demand for foodgrains as also for the agricultural raw materials for industry in the future. The area under cultivation, obviously, cannot be increased and the present 140 million hectares will have to meet the future increases in such demands. There is a strong reason for even a decline in the cultivated area because of the urbanisation and industrialisation, which in turn will exert much pressure on the existing, cropped area.

Science and technology have helped man to increase agricultural production from the natural resources like land. But the realisation that this has been achieved at the cost of the nature and environment, which support the human life itself, is becoming clear. It has been fully evident that the present pattern of economic development, which ignores the ecology and environment, cannot sustain the achievement of man without substantial erosion of the factors that support the life system of all living things on the Earth.

The evidence of the ill effects of development is well documented. As said earlier, we in India have to be concerned much more than any other nation of the world as agriculture is the source of livelihood of more than 6-7 million of our people and it is the foundation of the economic development of the country. There were times when people lived close to nature with access to flora and fauna in healthier and cleaner surroundings.

One has to look back at our present metropolitan cities or other large towns before the past fifty years as recorded in history/memmories of the present elder generation to see the striking differences in the surroundings in which the people lived there. Land, water and air, the most fundamental resources supporting the human life, have degraded into such an extent that they now constitute a threat to the livelihood of millions of people in the country. Ecological and environmental effects have been highly publicised all over the world.

Many times, these analysis have taken the shape of doomsday forecasts. Powerful interests in the developed western countries have also politicised these issues to take advantage of the poor nations of the world. Efforts to impose trade restrictions on the plea of environment protection are a direct result of these campaigns. But we have to recognise that the abysmal level to which we have degraded our resources, requires immediate remedial measures without terming the demand for them as the ploys of the rich nations to exploit the poor.

Another turn of the events has been the blame game for ecological problems stated at the Earth Summit and other international conferences. The developed countries, it is true, are to a great extent instrumental to degrade the environment. However, the poorer countries of the world including India cannot delay or ignore the need for remedial measures, which are to be effectively implemented. We cannot gloss over the fact that we have also contributed to the degradation of ecology; look at the droughts and floods, disappearance of forests, high noise level and air pollution in the cities which are our own creations.

Organically cultivated soils are relatively better attuned to withstand water stress and nutrient loss. Their potential to counter soil degradation is high and several experiments in arid areas reveal that organic farming may help to combat desertification. It is reported that about 70 hectares of desert in Egypt could be converted into fertile soil supporting livestock through organic and biodynamic practices.

India, which has some areas of semi-arid and arid nature, can benefit from the experiment. The organic agriculture movement in India received inspiration and assistance from IFOAM which has about 600 organisational members from 120 countries. All India Federation of Organic Farming (AIFOF) is a member of IFOAM and consists of a number of NGOs, farmers' organisations, promotional bodies and institutions. The national productivity of many of the cereal crops, millets, oilseeds, pulses and horticultural crops continues to be one of the lowest in the world in spite of the green revolution.

The fertiliser and pesticide consumption has increased manifold; but this trend has not been reflected in the crop productivity to that extent. The country's farming sector has started showing indications of reversing the rising productivity as against the increasing trend of input use. The unsustainability of Indian agriculture is caused by the modern farming methods which have badly affected/ damaged production resources and the environment.

The role of agriculture in economic development in an agrarian country like India is a predominant one. Agriculture provides food for more than 1 billion people and yields raw materials for agro-based industries. Agricultural exports earn foreign exchange. Modernisation of Indian agriculture began during the mid-sixties which resulted in the green revolution making the country a foodgrain surplus nation from a deficit one depending on food imports.

Modern agriculture is based on the use of high yielding varieties of seeds, chemical fertilisers, irrigation water, pesticides, etc., and also on the adoption of multiple cropping systems with the extension of area under cultivation. But it also put severe pressure on natural resources like, land and water. However, given the continuous growth of modern technology along with the intensive use of natural resources, many of them of non renewable, it is felt that agriculture cannot be sustainable in future because of the adverse changes being caused to the environment and the ecosystem.

The environmental non-degradable nature of the agricultural development and its ecological balance have been studied in relation to the modem Indian farming system by experts which shows exploitation of land and water for agriculture, and the excessive use of chemicals.

Chemical Contamination

Fertilisers

Consumption of chemical fertilisers (N,P,K) has been increasing in India during the past thirty years at a rate of almost half a million tonnes on an average, a year. It was only 13.13 kg/ha in 1970-71, 31.83 kg/ha in 1980-81 and 74.81 kg/ha in 1995-96. It shot up to about 96 kg/ha during 1999-2000. The present use of about 96 kg of fertilisers per ha in India appears to be modest compared to the advanced countries. Currently about 80 per cent of the fertiliser is consumed in only about 120 districts constituting less than 33 per cent of the gross cultivating area.

Experts point out that the efficiency of fertiliser use in India is only 30-35 per cent as the balance 65-70 per cent reaches the under ground water. The intensity of their use in a few regions and a few crops are causes of serious concern to human health, soil, water, environment and thus to the sustainability of agriculture production in the country. It is true that the increasing use of fertiliser at high rates has boosted agricultural production in the country. But it has also caused adverse impact on soil and water as well as environment.

Several studies on the effects of high level of fertiliser application on soil health have confirmed the adverse impacts. Both drinking and irrigation water wells in large numbers have been found contaminated with nitrates, some of them are having even 45 mg per litre, well above the safe level. Long term continuous use of high doses of chemical fertilisers badly affects the physical, chemical and biological properties of the soil.

A study at the University of Agricultural Sciences, Bangalore confirmed the deterioration of soil health because of the reduction in water holding capacity, soil pH, organic carbon content and the availability of trace elements such as zinc in case of ragi crop even with the application of normal doses of fertiliser in the long run. In the long run, increasing nitrogenous fertiliser use leads to the accumulation of nitrates in the soil. The application of sulphatic fertilisers leaves sulphates in the soil. Rainfall and excessive use of irrigation water cause these chemicals to change the alkaline or acidic nature of the soil.

The nitrates go to the rivers, wells, lakes etc. And also leak into the drainage system which goes into the drinking water contaminating the environment. It also causes depletion of the ozone layer adding to the global warming.

Use of nitrogen in the form of ammonium sulphate in the rice crop emanates ammonia polluting the atmosphere. The heavy metals present in the fertilisers and sewage sludge leach into ground water. Table 6 shows the content of some heavy metals in fertilisers and sludges.

Pesticides

The use of chemical pesticides began with the discovery of toxicological properties of DDT and HCH during the Second World War. Many chlorinated hydrocarbon insecticides like aldrin, dieldrin, toxaphane, chlordane, endosulfan, etc. came into the market during the second half of the last century. Simultaneously, organophosphate and carbonate compounds were employed in agriculture.

A new group of insecticides, such as premethrin, cypermethrin, fenalerate, etc. which were effective at low doses came into being in the 1970s. The use of pesticides has helped in increasing agriculture production and also led to the development of resistance in pests, contamination of the environment and resurgence of many pests. There are about 1000 agrochemicals in use in the world over.

India accounts for about 3.7 per cent of the total world consumption. At present, our consumption is about 90,000 tonnes of plant protection chemicals. It comes to about 500 grams per ha compared to 10-12 kg/ha in Japan and 5 kg/ha in Europe. However, the use of pesticides in India is uneven like the fertilisers. While in cotton it is about 3 to 4 kg/ha, in pulses it comes to below 500 grams/ha.

Pesticide application is also concentrated in some areas as in the case of fertilisers mentioned earlier. Agricultural chemicals have become a major input in Indian agriculture with the increasing demand for food, feed and fibre. The pesticide consumption was about 2000 tonnes annually during the 1950s. India happens to be the second largest manufacturer of pesticides in Asia after Japan. It is also of interest to know that in spite of increased consumption of plant protection chemicals, the produce loss due to insects and pests increased by 5 times during the period from 1988 to 1995.

Increasing application of fertiliser also leads to increasing use of pesticides to control pests and diseases. The trend of increasing fertiliser use also compels the farmers to enhance the use of pesticides as well. For example, the use of fertilisers in increasing amounts leads to growth of weeds and in the process of weedicide use many plants growing nearby also get killed, which reduces the biodiversity. Meanwhile, the weeds also develop resistance to herbicides and the quest to formulate even powerful herbicides begins.

Consumption of pesticides increased from 24.32 thousand tonnes in 1970-71 to 75 thousand tonnes in 1990-91 and it slowed down during the subsequent period. Insects, pests and diseases like viral, bacterial and fungal affect the high yielding varieties of crops. Almost all pesticides are toxic in nature and pollute the environment leading to grave damage to ecology and human life itself. This indiscriminate use leaves toxic residues in foodgrans, fodder, vegetables, meat, milk, milk products, etc. besides in soil and water.

High doses of pesticides severely affect the aquatic animals, fish and the wild life. Insects develop resistance to insecticides in crops like cotton and in turn force the farmers to the excessive use of them. Cases of pesticide poisoning and human and animal deaths are also reported. Pesticides irritate the skin and the respiratory system in the humans gets damaged. It was found that all water bodies like, rivers, canals, lakes, tanks and ponds and also the costal water were contaminated with high amounts of DDT, HCH and other organochlorine pesticides. River water is seen as more contaminated than other water sources.

Contamination of drinking water with DDT and HCH is reported from different states. Since the concentrations of contaminants are higher than MRL (0.5 ppb) values fixed by the Environmental Protection Agency (EPA), the seriousness of the problem can be gauged. Pesticides also contaminate animal feeds and fodder. Green fodder, paddy and wheat straw contain residues of DDT and HCH. Several studies have confirmed this trend.

Milk and milk products are also affected by the pesticide use. Both bovine and human milk showed high levels of pesticide contamination. The sources of contamination of bovine milk are traced to the fodder and feed concentrates and in case of the human milk, the consumption of contaminated food by the lactating mothers is reported to be the reason. Infant formula/baby milk powders also showed DDT and HCH contamination level ranging from 94 to 100 per cent. Butter and ghee, the other animal products revealed high contamination levels in many parts of the country.

Cereals like wheat and rice were seen contaminated highly by pesticides like, DDT, HCH and malathion. The case of vegetable, vegetable oil, honey, fish etc. is also not different as they too have unacceptable high pesticide residue content levels. The adverse effect of pesticide contamination on humans in India is understood from the study of dietary intake. Such studies, although a few in number, have confirmed high levels of pesticides contamination which come to more than 3 to 5 times than the agriculturally developed countries.

The daily intake of pesticide per individual is estimated to be about 0.51 milli-grams which is above the accepted level. The Indian Institute of Horticulture Research has reported contamination of 50 per cent of the fruits and vegetables sold in the Bangalore market with the residues of DDT and HCH. Use of herbicides over a period results in the shift of the weed flora. The weeds of minor importance, often, become major weeds. Repeated application of weedicides helps the development of resistance in weed at alarming proportions.

The remedy recommended is rotation of herbicides or the use of other herbicides. Any way, the end result is contamination of ground water and soils inflicting damage on environment. The number of herbicides registered in India comes to about 28 in 1997-98 which was only 10 before 10 years. This is often compared to about 300 herbicides available in the North America. There are only 10 herbicides manufactured in the country and the herbicides consumption was about 6000 Tonnes during 1994-95. It is reported that the level of herbicide use in rice, wheat, and tea in India is almost the same that of the world at large.

Sugarcane, soyabean, groundnut, coffee, cotton, onion and potato are the other crops, which find widespread application of herbicides in India. The contamination of water, air and soil with toxic synthetic fertilisers, pesticides and herbicides leads to increasing deaths of many creatures, and to human illness and mortality. The end result is loss of biodiversity and natural harmony, increased expenditure to purify water, air, etc. The toxins in the food crops cannot be removed and the threat to human existence itself seems to be real. The firms engaged in the manufacture and supply of agricultural inputs have a vested interest in keeping the input use increasing. Besides, they influence the government policy towards agriculture.

Salinity and water logging

Water is one of the important inputs for the vigorous growth and high yields of crops. The modernisation of Indian agriculture has resulted in the increased use of irrigation water. The area under irrigation has grown substantially during the past three decades. The gross irrigated area of 38.18 million ha in 1970-71 increased to 49.73 million ha in 1980-81 and the next decade ending

1990-91 saw this further rising substantially to about 62.47 million ha. It increased to 72.78 million ha during 1997-98.

Heavy irrigation is necessary to get high production, as the new varieties cannot withstand water scarcity. This leads to salinity and water logging leaving the land uncultivable. Over exploitation of underground water is another effect. When water table falls, increasing energy will be required to lift water for irrigation. Irrigation is necessary for the vigorous growth and high yields of crops in the modern method of cultivation.

Many of the crops, particularly the rice and wheat high yielding varieties need more irrigation water than the traditional varieties. The area under irrigation in the country is only about 35 per cent and the remaining is still dependent on rains. So, there is a necessity to use irrigation water judiciously. Its excessive use results in severe ecological dangers like water logging of vast cultivated areas by seepages from canals.

The loss of water through seepages and evaporation is estimated to be about 38 per cent. Flooding also results in run off and leaching losses of fertiliser nutrients, pesticides and soil particles. Excessive use of canal water makes the field vulnerable to soil erosion. The excessive irrigation in certain areas results in wastage as evident from the water logging of vast cultivated areas caused by the seepage from many virater sources. Water logging is harmful to the soil. Seepage of canal water leads to salts present in the lowest layer of soil come up to the surface and the soil may turn alkaline or saline.

Dams and multipurpose projects degrade the soil in the command area due to soil salinity and water logging. The chambel region in Rajasthan and Madhya Pradesh, the command area of the Bhakra Nangal, etc are the examples of water logging created by huge water irrigation projects. Crops irrigated by sewage water have adverse effects on the health of the human population consuming the produce. The workers work on these farms also face health hazards.

Depletion of energy resources

Chemical fertilizers, pesticides, herbicides, etc are manufactured using the non-renewable materials like the fossil fuels. The global demand for oil and natural gas is increasing and thus the price of the inputs to agriculture is bound to rise. India's petroleum resources, which presently meet only about 30-35 per cent of the consumption demand, are under pressure. Increasing demand for chemicals and energy in agriculture sector will have affects on our energy sources.

The investments in agriculture have to be increased to meet the rising input costs and larger areas are brought under farming to earn profits. Large farms have to transport the produce to distant areas. Again, energy will be required for transportation, processing and packaging. The rice-wheat cropping pattern and the cultivation of crops like sugarcane require high irrigation, which results in the depletion of water level. Singh and Singh found that the water level in the states of Punjab and Haryana had gone down by 0.3 to 1 m per annum during a period of 10 years due to the excessive use of water for paddy crop.

Input-output imbalance

A crop, in its growth process, incorporates a part of the soil fertility into the parts of the plant. The roots remain in the soil. The leaves and stems are fed to the cattle/burnt as fuel/directly returned to the soil. The consumed part by cattle and human also go back to the soil. The practice of commercial farming leads to continuous export of the soil fertility to outside the farming areas as the organic matter leaves the locality.

The soil nutrients in the form of farm produces continue to be exported. The import of chemical fertilisers cannot compensate the loss of soil nutrients through exports. The soil becomes powdery and gets eroded by wind or rain. If the harvests are exported from the country, the loss is higher.

The Earth can produce only a limited amount of biomass from a given area. If man tries to extracj: more, the system degenerates. If we realise that we live in a closed eco-system and act in tune, the resources can be used iri a sustainable way. The maintenance of a dynamic equilibrium balances the inputs and outputs in a region. If agriculture becomes unsustainable, no society can survive for a long time.

Cultivated area

Not only the intensive cultivation through the use of technological inputs, but also the extensive crop production through Increase in the area under cultivation has been an important aspect of modern agriculture seen in India. Increasingly areas under forests are brought under plough along with the marginal, sub-marginal and undulating land. The net sown area was 140 million ha in 1970-71 and stood at 142 million ha at the end of 1997-98.

Genetic diversity

The genetic base of crops is very important and a reduction of genetic diversity leads to the emergence of pests on a large scale. Farmers, in olden times, apart from' using the crop rotation methods to maintain the soil fertility also relied on the genetic means to increase crop production. Relying exclusively on nation's own reserves of fertility and immunology, the farming community by evolving trial and error methods discovered hybrid varieties of crops by crossing the related strains.

These crosses were from the same environment and no violence was used to separate them from nature by maintaining the ecological balance. The high yielding varieties of crops are the crosses from different environments and distantly related strains. For example, the high yielding rice variety got by the crossing between the dwarf and non-dwarf varieties has major genetic weaknesses.

The dwarf gene is susceptible to pest and viral attacks and the seed cannot manifest its potential without chemical fertiliser. Thus, an artificial environment has to be created for the growth of the crop. Thus synthetic fertilisers supplant natural fertility, which results in larger population of pests. The new technology adopted then depends upon the replacement of the local/traditional varieties of seeds. But this results in the reduction of genetic diversity and increase in genetic erosion. These modern technologies are but the result of clever manipulations of nature's genes.

Low productivity

The productivity of cereals, millets, oilseeds, pulses and plantation crops is very low in comparison with those in other countries in the world. This is in spite of our success in improving the quality of seeds and adoption of efficient technology. The impact of green revolution is showing signs of weakness and production appears to have decreased even after an increase in the inputs used.

The production of foodgrains in the country increased very substantially during 1960 to 1980 to reach 160 million Tonnes from 60 million tonnes. But the decade ending 1990 and 2000 did not witness such increases and the attainment of the targeted production of 240 million tonnes to meet the demand of the population by 2010 seems to be difficult. The reasons attributed to the low productivity are the drastic reduction in soil nutrients in the areas where fertiliser is used intensively in which the organic matter is not supplemented.

Benefits of organic farming

Organic agricultural practices are based on a maximum harmonious relationship with nature aiming at the non-destruction of the environment. The developed nations of the world are concerned about the spreading contamination of poisonous chemicals in food, feed, fodder and fibre. Naturally, organic farming system is looked upon as one of the means to remedy these maladies there.

However, the major problem in India is the poor productivity of our soils because of the low level content of the organic matter. The efficiency of the organic inputs in the promotion of productivity depends on the organic contents of the soil. There were many resemblances of organic farming principles in the traditional agriculture of India. But the former gives a more open and verifiable scientific foundation than the latter.

Healthy foods

A study conducted in USA on the nutritional values of both organic and conventional foods found that consumption of the former is healthier. Apples, pears, potatoes, corn, wheat and baby foods were analysed to find out 'bad' elements such as aluminum, cadimum, lead and mercury and also 'good' elements like boron, calcium, iron, magnesium sellenium and zinc. The organic food, in general, had more than 20 per cent less of the bad elements and about 100 per cent more of the good elements.

Soil quality

Soil quality is the foundation on which organic farming is based. Efforts are directed to build and maintain the soil fertility through the farming practices. Multicropping, crop rotations, organic manures and pesticides, and minimum tillage are the methods employed for the purpose. Natural plant nutrients from green manures, farmyard manures, composts and plant residues build organic content in the soil.

It is reported that soil under organic farming conditions had lower bulk density, higher water holding capacity, higher microbial biomass carbon and nitrogen and higher soil respiration activities compared to the conventional farms. This indicates that sufficiently higher amounts of nutrients are made available to the crops due to enhanced microbial activity under organic farming.

Crop productivity and income

Field trials of organic cotton at Nagpur revealed that during the conversion period, cotton yield was low compared to the conventional and integrated crop management. However, the yields of organic cotton started rising from third year. Cotton yields under organic, conventional and the mixed systems were 898, 623 and 710 kg/ha respectively at the end of the fourth year of the cultivation. The yield of soyabean under organic farming was also the highest compared to the other two systems.

The Central Institute for Cotton Research, Nagpur conducted a study of economics of cotton cultivation in Yavatmal district of Maharashta. The cost of cultivation of cotton was lower in the organic farming than in the modern system. The low costs were due to the non-use of fertilisers and chemical insecticides. As a result of the low yields during the conversion period, the net income from the organic farm was lesser than the conventional farm. But the yield under organic method increased progressively equalling it to that of the conventional system by the sixth year.

The input costs were low under organic farming and with a 20 per cent of premium prices of output, the net income increased progressively from fourth year under organic fanning. The appreciation of net income from organic cotton cultivation by the sixth year was 80 per cent over the conventional crop. Results reported from 1050 field demonstration cum trials under the National Project on Development and Use of Biofertilisers in different parts of the country show an increase of 4 per cent in yield in plantation crops, 7 per cent in fruit crops, 9 per cent in wheat and sugarcane, 10 per cent in millet and vegetable, 11 per cent in fibre, condiments and spice crops, 14 per cent in oilseeds and flowers and 15 per cent in tobacco.

A study of 100 farmers in Himachal Pradesh during a period of 3 years found that the total cost of production of maize and wheat was lower under organic farming and the net income was 2 to 3 times higher. Both productivity and premium prices contributed to the increased profitability. Another study of 100 farmers of organic and conventional methods in five districts of Karnataka indicated that the cost of organic farming was lower by 80 per cent than that of the conventional one. The cost benefit ratios mentioned for various crops were:

Crops	Organic	Inorganic
Groundnut	1:1.26	1:1.31
Jowar	1:1.36	1:1.28
Cotton	1:1.34	1:1.24
Coconut	1:1.70	1:1.31
Banana	1:3.66	1:2.82

Low incidence of pests

The study of the effectiveness of organic cotton cultivation on pests at the farm of Central Institute for Cotton Research, Nagpur revealed that the mean monthly counts of eggs, larva and adults of American BoUworm were far lesser under organic farming than under the conventional method. Bio-control methods like the neem based pesticides to Ti-ichoderma are available in the country.

Indigenous technological products such as Panchagavya which was experimented at the University of Agricultural Sciences, Bangalore found to control effectively wilt disease in tomato.

Employment opportunities

According to many studies, organic farming requires more labour input than the conventional farming system. Thus, India which has a very large amount of labour unemployment and under employment will find organic farming an attraction. Moreover, the problem of periodical unemployment will also get mitigated because of the diversification of the crops with their different planting and harvesting schedules resulting in the requirement of a relatively high labour input.

Indirect benefits

Several indirect benefits from organic farming are available to both the farmers and consumers. While the consumers get healthy foods with better palatability and taste and nutritive values, the farmers are indirectly benefited from healthy soils and farm production environment. Eco-tourism is increasingly becoming popular and organic farms have turned into such favourite spots in countries like Italy. Protection of the ecosystem, flora, fauna and increased biodiversity and the resulting benefits to all human and living things are great advantages of organic farming which are yet to be properly accounted for.

Proposed objectives of organic farming

The broad objectives of organic farming in India can be the following in the light of the discussions on the adverse effects of the conventional farming system the country practiced for about 30-40 years and the potential benefits of the organic methods:

— Sustainable agriculture
— Increasing agriculture production
— Food self-sufficiency
— Environmental protection
— Conservation of natural resources
— Rural development

REFERENCES

Anon, *Green Revolution to Organic Revolution, Comprehensive Agric-Commodity Intelligence,* April, 2003, pp. 21-35.

Dahama, AK, *Organic Farming for Sustainable Agriculture,* Agribios (India), Jodhpur, 2002.

Deb, DL and Joshi, HC, "Environment Pollution and Agriculture," in *Natural Resource Management for Sustainable Agriculture and Environment,* Angkor Publications (P) Ltd., New Delhi, 1994, pp. 227-293.

Prakash, TN, "A Theoretical Framework To Promote Organic Produce Marketing in India," *Indian Journal of Agriculture Marketing,* Conference Number Special, 2003, pp. 1-16.

7

Ecosystem Approach to Organic Food Production

Current approaches to ecological agriculture such as integrated pest management, integrated plant nutrition systems and conservation tillage, consider only one aspect of the farming system components: pest ecology, plant ecology and soil ecology, respectively. The ability to define strategies that combine these and other management elements into a single approach is proper to organic agriculture. Organic management focuses on the food web relations and element cycling and aims to maximise the agro-ecosystem's stability and homeostasis.

Organic agriculture, through its systemic approach and avoidance of agro-chemicals, prevents natural resource degradation and the loss of land and productive potential. In organic agriculture, nature is both instrument and aim. As organic farmers cannot use synthetic substances they need to restore the natural ecological balance because ecosystem functions are their main productive "input". For example:

— Many unspecific pests like aphids, thrips, whiteflies or spider mites, economically damaging in many crops, can be kept below the economic threshold with naturally occurring or purposely released predators and parasitoids. The former are direct goods and services of hedges, botanically diverse field margins, intercropping or weedy undergrowth, the latter do better when released in botanically and ecologically enriched habitats.

— The only way to suppress soil-borne pests and diseases in organic agriculture are wide crop rotations with several components of botanically different crops. Adhering to such rotations is crucial to providing agro-ecosystem diversity.

— Diverse crop rotations, or agro-forestry systems, guarantee a better uptake of nutrient elements from the soil and very efficient use of water and light, thanks to varying spatial and temporal root growth and leaf dispersion.

— Soils with a high functional diversity of microorganisms, which occur very often after decades of organic agriculture practice, develop disease suppressive properties and can help to induce resistance in plants.

— By restricting farm inputs farmers have to use preventive techniques appropriately. The ban on herbicides, for instance, makes it impossible to ignore good crop rotation principles and disastrous in terms of yields and long-term problems with weeds. The ban on soluble and purchased fertilisers makes nutrient-conserving crop rotations and the sparing use of organic fertilisers to reduce losses, economically worthwhile.

As organic farmers cannot use synthetic substances (e.g. fertilisers, pesticides, pharmaceuticals) they need to restore the natural ecological balance because ecosystem functions are their main productive "input". The direct economic value of ecosystem functions for organic farmers is an excellent example of "benefit-sharing". To master ecosystem functions in order to produce decent high-quality yields, adaptive management practices are crucial:

growing conditions in organic agriculture are neither optimal, static nor predictable.

Farmers must become excellent observers and must be trained to react flexibly, often intuitively and appropriately to the local context. In contrast, other farming methods try hard to condition the environment of the plants for optimal growing conditions by using a wide range of inputs. Moreover, most recommendations and fertilisation and spraying programmes are highly standardised and not site-specific. Organic agriculture is an approach, which is not confined to production alone. Intersectoral cooperation between farmers, environmentalists, processors and traders, quality controllers and consumers is a very common characteristic of organic agriculture.

ORGANIC AGRICULTURE AND ABIOTIC RESOURCES

Soil Fertility

As the central basis for all agricultural activity, soil is one of the chief natural resources. Since organic farmers cannot compensate for a loss in soil fertility by inputs of synthetic nutrients, the building and maintenance of soil fertility is a central objective of organic agriculture. The impact of organic agriculture on soil properties has been covered extensively by research with a special focus on the relevant parameters of organic matter content, biological activity and soil erosion.

Organic Matter Content

The environmental importance of organic matter content is its capacity to limit physical damage and to improve nutrient availability as well as biological activity. Research on organic matter concentrates on measuring the soil organic carbon content parameter.

Fertilisation in organic agriculture is based on organic substances such as farmyard manure, compost, green manure, plant residues and commercial organic N-fertilisers. Consequently, there is an extensive supply of organic matter passing through aerobic decomposition processes.

Mineralisation and decomposition processes are influenced by humidity, temperature and oxygen. Under humid tropical conditions these processes run faster and all year round, whereas under temperate conditions they are slower and come to a halt during the colder months. Soil type also plays a role. Sandy soils dry out quickly, slowing down the decomposition process, ferralitic soils on the other hand are generally not very fertile, but they encourage fast decomposition and the building-up of stable organic matter.

Soil Micro-Organisms

Intense biological activity promotes metabolism between soil and plants and should be a focus of sustainable plant production and fertiliser management. In contrast to conventional agriculture, organic farmers depend more on a high and sustained supply of organic substances including crop rotations with clover/grass ley, underseeds, catch crops, green and animal manure. Under European conditions, organic agriculture performs better than conventional agriculture with regard to certain parameters: for example, 30 to 100 percent higher microbial activity and a significantly higher biomass (+30 to 40 percent), density (+ 50 to 80 percent) and species diversity of earthworms, a key soil-macro faunal species.

Soil is a habitat for plants, animals and microorganisms. As plants build up organic matter, soil animals feed on them and their debris, whilst microbes decompose the complex organic compounds to their mineral components and to carbon dioxide. A living soil is central to soil fertility because it is the activity of soil organisms that makes available the elements in plant residues and organic debris entering the soil. Part of this material, however, remains in the soil and contributes to its stabilisation by humus build-up.

Various research results show that the activity of microorganisms is higher in organically- than in conventionally-managed soil. As a consequence, in organically managed soils, nutrients are recycled faster and soil structure is improved. For example, Fliessbach et al found in Switzerland up to 90 percent higher total mass of microorganisms in organically managed soils. As for soil fungi, Elmholt found a higher number and abundance of saprophytic soil fungi with a higher potential of decomposition of organic material. Mycorrhizae, important representatives of soil fungi, live as symbionts with plant roots. The degree of mycorrhizal root colonisation was found to be distinctly higher in organic plots as compared to conventional plots.

As mineralisation processes proceed much faster on ferralitic soils in the tropics and subtropics than on soils typical to temperate and continental zones, a high organic matter content and high biological activity are the prerequisite for sustainable soil fertility.

Soil Erosion

Soil erosion is assumed to be the main cause of soil degradation around the world. The loss of fertile top soil by erosion results in a lower yield capacity on the onehand and in a undesired transfer of nutrients, pesticides and sediments in surface water on the other.

Reganold et al found during a long-term comparison of soils from selected conventional and organic farms of the same soil type near Spokane in Washington, USA, that topsoil was 16 cm thicker on organically managed plots. This was most probably due to inclusion of a green manure legume crop in the third year of rotation and fewer tillage operations on the organic field. The organically-farmed soil not only had deeper topsoil but also significantly higher organic matter content and suffered less soil erosion than the conventionally farmed soil.

Research Results on Soil and Organic Agriculture

A thorough comparison of relevant soil parameters on conventionally- and organically-managed soils is provided by the long-term DOC-tria carried out by the Swiss Research Institute of Organic Farming (FiBL). The results of the DOC trial can be summarised as follows:

— Organic matter content is usually higher in organically-managed soils than in exclusively minerally-fertilised conventionally-managed ones, thanks to organic fertilisation methods. High organic matter content also helps to avoid soil acidification.

— Organic soil management improves soil structure by increasing soil activity, thus reducing the risk of erosion.

— Organic crops profit from root symbioses and are better able to exploit the soil.

— Organically farmed soils have significantly higher biological activity (earthworms, fungi, bacteria, microorganisms) than conventionally-managed ones. Thus nutrients are transformed more rapidly and soil structure is improved.

— Organic management promotes the development of soil fauna such as earthworms and above ground arthropods, improving the growth conditions of the crop. More abundant predators help to control harmful organisms.

Organic soil management improves soil structure by increasing soil activity, thus reducing the risk of erosion.

Ground and Surface Water

The detrimental effects of intensive agriculture on ground and surface water are largely due to erosion and to nitrate and pesticide pollution. The chief threats to water quality posed by agriculture are: high organic fertilisation levels in combination with high stocking rates, the excessive application of mineral N-fertilisers; the lack of a protective soil cover; narrow crop rotations and frequent tillage; high levels of available nitrogen after harvest, and contamination of water with synthetic pesticides.

As organic agriculture uses no synthetic pesticides, there is no risk of synthetic pesticide pollution of ground and surface waters. The reasons for the lower nitrate leaching rates in organic agriculture are the ban on mineral N-fertilisers and lower livestock densities. These constraints set up by the organic agriculture standards lead to nitrogen on organic farms nitrogen being—in economic terms—a quantitatively scarce factor.

The economic consequences of nitrogen being scarce on organic farms are quite impressive: the opportunity costs (cost of producing on-farm) of 1 kg nitrogen on organic farms can amount to seven to sixteen times the cost of mineral N-fertilisers. So it is not surprising that in contrast to conventional farms where manure and slurry are often a waste problem, organic farmers are forced to develop efficient nitrogen management strategies like intercropping, catch cropping, optimal ploughing-in of legume crops or limiting the use of liquid manure to avoid nitrogen losses.

As a result, in some regions of Germany and France, waterworks subsidise conversion to organic agriculture in water protection areas as an economically efficient solution for reducing the cost of cleaning-up drinking water by minimising the nitrate and pesticide contamination of groundwater. Waterworks subsidise conversion to organic agriculture in water protection areas as an economically efficient solution for reducing the cost of cleaning-up drinking water by minimising the nitrate and pesticide contamination of groundwater.

While scientific results from other climatic zones are scarce, positive effects of organic agriculture on the nitrate leaching risk can be reported from a citrus farm in Cuba. Under organic fertilisation management based on compost with 60 kg available N per ha, the farm achieved exactly the same yield level as under conventional fertilisation management with 200 kg of mineral N.

Nutrient Use

An adequate and balanced supply of nutrients in the soil is essential for several reasons. Nutrient surpluses might result in nutrient losses which could subsequently lead to water and air contamination and eutrophication. However, nutrient deficiency is synonymous with the over-exploitation of soil nutrients in the long run, which consequently leads to a decrease in yield and crop quality.

Research by Freyer in Switzerland shows that only 14 percent of all organic farms have an N-surplus, and only 1.5 percent had a P-surplus. Most of the organic farms have a negative N- and P-balance. Results from different European countries comparing phosphorous and potassium balances of conventional and organic farms are presented in Table 1: even though the ranges from study to study vary a great deal, it can be concluded that the phosphorous and potassium surpluses of organic farms are significantly lower than on conventional ones.

Due to negative nutrient balances as shown in Table 2, the question arises whether organic agriculture methods cause gradual loss of soil minerals. First of all, the proportion of soluble nutrients is lower on organically-managed soils. On the other hand, Mäder et al found no decrease in organic yields as an indicator for nutrient deficiency on farms which are managed organically for more than 30 years.

Table 1: Examples for P, K balances (kg/ha) comparing organic with conventional farms from different European countries

	P balance (kg/ha)		K balance (kg/ha)	
	Organic	Conventional	Organic	Conventional
Sweden	-12	+37	-4	+39
Netherlands				
Cash crop farm	+18	+23	+31	+25
Horticulture	+32	+60	+119	+110
Dairy farm	+8	+31	-	-
Germany				
Mixed farm	-4	+13	-27	+31
Dairy farm*	-2	+5	+7	+20

Energy Consumption

Energy consumption in agriculture includes the direct consumption of fossil energy (e.g. fuel and oil), as well as indirect energy consumption (e.g. from the production of synthetic fertilisers and pesticides). Leaving aside indirect agricultural energy consumption, OECD statistics indicate that agriculture contributes only 2 percent to total direct energy used in OECD countries. Nevertheless, limited fossil energy resources and the climatic relevance of its use require efficient energy use, even in agriculture. Relevant parameters for evaluating energy use in agriculture are energy consumption and energy efficiency.

Considering both, direct and indirect energy consumption, scientific calculations on energy consumption per hectare indicate that organic farms use less energy than conventional farms: several researchers calculated the energy consumption of organic farms to amount 64 percent of that on conventional farms.

The determining factor for energy consumption of a specific crop is cropping management, which includes tillage intensity, manuring and weed control. On a per hectare scale, all authors determined lower energy consumption on organic farms. For organic potatoes and apples, however, energy consumption per output unit is higher relative to conventional production. This is the result of higher energy input for mechanical measures like weed control in organic production and of lower mineral N-fertilisation use in conventional production.

A second parameter appropriate for evaluating energy use is energy efficiency. It provides information about the ratio of energy input and output. Comparing rotations of different production systems in Iran, Zarea et al. found the energy efficiency of organic agriculture to be 81 percent better compared to high-input conventional agriculture. In a similar investigation in Poland Kus & Stalenga calculated a 35 percent higher energy efficiency of organic compared to conventional agriculture. Under Mediterranean conditions in Italy, a 25 percent higher efficiency in organic wheat and an 81 percent higher efficiency in organic vineyard production systems were found.

Even though the ban on synthetic pesticides might lead to higher fuel consumption on organic farms due to increased mechanical weed control, research results presented below show that with respect to energy consumption, organic agriculture is performing better than conventional agriculture. The main reasons for this are:

— No input of mineral N-fertilisers which require a high energy consumption for production and transport on organic farms;

— Lower use of highly energy-consumptive foodstuffs (concentrates);

— Lower input of mineral fertilisers (P, K);

— Ban on synthetic pesticides.

With respect to energy consumption, organic agriculture is performing better than conventional agriculture. However, higher labour requirements may be expected on organic farms under European conditions due to a higher share in the production of labour intensive crops (e.g. vegetables) and on-farm marketing and processing activities. Arable crops also demand greater labour inputs, for example, for mechanical weeding.

Most commonly, labour use per hectare of utilised agricultural area is on average 10 -20 percent higher on organic farms. In a European context, the labour input is higher on organic arable and mixed farms, while organic dairy farms use the same amount of labour, or less, than comparable conventional farms. On horticultural farms, labour requirements are much higher than on conventional farms. Few data exist on pig and poultry farms, but labour per hectare of utilised agricultural area seems to be similar to conventional farms, as livestock density is reduced.

ORGANIC AGRICULTURE AND BIODIVERSITY

For hundreds of years, agriculture has contributed substantially to the diversity of species and habitats, and agriculture has formed many of today's landscapes. Over the last century, however, modern intensive agriculture, with its high input of synthetic pesticides and fertilisers and monocrop specialisation, has been detrimental to the diversity of genetic resources of crop varieties and livestock breeds, to the diversity of wild flora and fauna species and to the diversity of ecosystems. The 2000 IUCN Red List of threatened species highlights habitat loss as the main threat to biodiversity, with agricultural activities affecting 70 percent of all threatened bird species and 49 percent of all plant species.

Because of the increasing loss of agricultural biodiversity on a global scale, the Convention on Biological Diversity has developed a work programme on this subject in 1996. This programme states inter alia that farming practices that stop degradation while restoring and enhancing biological diversity should be encouraged, including organic agriculture.

Organic agriculture is dependent upon stabilising agro-ecosystems, maintaining ecological balances, developing biological processes to their optimum, and linking agricultural activities with the conservation of biodiversity. Wild species perform a variety of ecological services within organic systems: pollination, pest control, maintenance of soil fertility. Thus, higher levels of biodiversity can strengthen functions essential for farming systems and therefore, agricultural performance. Enhancing functional biodiversity is a key ecological strategy to bring sustainability to production on organic farms.

Organic systems also use substantially fewer external inputs and do not use synthetic chemical fertilisers, pesticides, genetic modified organisms and pharmaceuticals. Instead, systems are designed to work in harmony with nature in order to determine agricultural yields and disease resistance. By respecting the natural capacity of plants, animals and the landscape, organic agriculture aims to optimise quality in all aspects of agriculture and the environment.

Biological pest control on organic farms, for example, relies on maintaining healthy population of pest predators and parasitoids. A study in California comparing conventional with organic tomato fields showed a higher abundance of natural enemies and greater richness of species in organic tomato fields. There was no significant difference for any type of damage to tomato foliage or fruit, showing that the organic system achieves the same levels of pest control without having to apply synthetic chemical pesticides.

Organic agriculture is thus committed to the conservation and enhancement of biodiversity within agricultural systems, both from a philosophical perspective and from the practical viewpoint of maintaining productivity. To this end, the importance of biodiversity as part of a well-balanced organic system is part of the IFOAM International Basic Standards for Organic Agriculture Production and Processing. Numerous scientific studies, mainly from Europe and North America, give evidence that on organic farms biodiversity is higher than that on conventional and farms. Biodiversity is generally assessed at three distinct levels:

— *Genetic diversity*: the variation between individuals and between population within a species;
— *Species diversity*: the different types of plants, animals and other life forms within a region or community;
— *Ecosystem diversity*: the variety of habitats found within an area (grassland, marsh, and woodland for instance).

In general, the degree of biodiversity in agro-ecosystems depends on four main characteristics of agro-ecosystems:

— The diversity of vegetation within and around the agro-ecosystem.
— The permanence of various crops within the agro-ecosystem.
— The intensity of management/farming.
— The extent of isolation of the agro-ecosystem from natural vegetation.

DIVERSITY OF GENETIC RESOURCES IN ORGANIC AGRICULTURE

Higher Crop Diversity

Since organic farms are mostly mixed farms, integrating animal husbandry with crop production, using vast and diverse rotations, intercrops and green cover crops, and maintaining soil fertility by cultivating nitrogen fixing legumes, they display a higher diversity of domesticated species than conventional farms.

Hausheer et al evaluated crop rotations on 110 organic, integrated and conventional farms in a Swiss pilot farm project. They found more diverse crop rotations (4.5 different crops in organic as opposed to 3.4 different crops in integrated farming) and a higher number of crops, including perennials, vegetables and herbs (10.2 in organic and integrated farms; 7.4 in conventional farms).

Maintenance of Genetic Resources

Today, the adoption of high-yielding, uniform cultivars and varieties has led to a considerable reduction in the number of plants and animals used in agriculture. Only 120 cultivated plant species and 14 mammalian and avian species provide 90 percent of human food supply.

Evidence of the trend towards monoculture and uniformity is given by the fact that in India, under the Green Revolution, the number of cultivated rice varieties has decreased from more than 100 000 to 10; also, 50 percent of the goat breeds, 20 percent of the cattle breeds and 30 percent of the sheep breeds are in danger of extinction. In Mexico, only 20 percent of the maize varieties reported in 1930 are now known. In China, nearly 10 000 wheat varieties were used in production in 1949; by the 1970s, only about 1000 remained in use. This trend is just as visible in animals: 740 animal breeds became extinct during the twentieth century. Currently, 1350 breeds face extinction, with two breeds being lost each week.

Organic producers look for productive varieties suited to their local climatic and soil conditions and that are not susceptible to disease and pest attack. Organic producers look for productive varieties suited to their local climatic and soil conditions and that are not susceptible to disease and pest attack. Organic agriculture standards recommend the cultivation of site adapted crop varieties, characteristics often found in the older native cultivars. This, however, does not necessarily mean that organic agriculture sets narrow limits to the use of modern maximum yield varieties, which are often chosen for pest/disease resistance purposes. Still, the preservation of native varieties and breeds is an important initiative of the organic movement, but their actual use depends on individual farmers.

Many seed banks and indigenous variety conservation programmes worldwide are linked to organic agriculture projects. For instance, the Sustainable Agriculture and Rural Development Project in Kenya is working with communities in the Gilgil district to develop organic systems to increase food security through a community indigenous seed conservation programme. Indigenous seeds have been shown to perform better in the harsh drought conditions.

Genetic Engineering and Agricultural Biodiversity

Apart from the adoption of high-yielding, uniform cultivars, a further possible threat to genetic diversity, and biodiversity in general, are the side-effects of the release of genetically engineered organisms into the environment. Genetically engineered plants designed to control pests can have negative side-effects on beneficial insects and further non-target organisms as well. Oilseed rape with genetically induced resistance to insects has been reported to damage beneficial insects such as honey bees. The use of herbicide resistant plants can result in greater use of herbicides, increasing the negative effects of intensive farming on natural biodiversity.

Recently, it was reported that transgenic DNA introgressed into traditional maize landraces in Oaxaca, Mexico. Concerns have been raised about the potential effects of transgenic introductions on the genetic diversity of crop landraces and wild relatives in areas of crop origin and diversification, as this diversity is considered essential for global food security.

Insofar as organic agriculture is dependent upon maintaining ecological balances and diverse agro-ecosystems, genetic engineering is a contradiction to the principal aims of organic agriculture. Organic agriculture does not allow genetic engineering in its standards as genetic engineering focuses on genetic makeup without taking into account the complete organism or farming system in which the organism functions.

Species Diversity

Floral diversity has been shown to contribute to ecosystem stability, while the invertebrate community associated with field boundaries performs many ecosystem functions including biological control of pests and diseases, pollination and food resource for higher trophic levels.

Floral Diversity

Today, the diversity of the typical wild flora on arable fields, which is the main habitat for a wide range of species, is at risk. Many species are endangered due to agricultural intensification, including the intensive use of mineral fertilisers and of herbicides, intensive soil management and the destruction of habitats. In grassland species, diversity is also decreasing due to intensification of grazing management and to high inputs of fertilisers.

Whereas in conventional agriculture weeds are considered competitive to the crop and are eliminated by herbicides, in organic systems some of the accompanying plants are desired to a certain degree and considered useful, as they provide a wide range of ecological services. These services include protection from soil erosion, providing shelter and alternative food resource for beneficial organisms and pollinators.

In all cases, organic agriculture showed higher floral diversity. Several comparative analyses carried out in Europe have shown plant species diversity in organic arable fields to be 30 to 350 percent higher than in conventional ones. In field edges, the number of plant species can be twice as high as in conventional fields, and the inner parts of the fields can have up to six times as many species compared to conventional fields. In Sweden a number of declining, endangered and rare species were recorded on organic arable fields, showing that organic agriculture can contribute to maintaining biodiversity. In Romania, a weed survey in maize fields showed a significantly higher diversity of weed species in the organic maize field.

In most cases, organic agriculture conserves better than conventional agriculture the site-typical plant community of floral species of arable land. A survey comparing organic with conventional fields showed that in the organic system the share of the fields with endangered floral species was 79 percent as opposed to 81 percent 27 years earlier, showing that the share had almost not changed. In the conventional fields the rate had dropped from 61 percent to 29 percent. The higher floral diversity and abundance in organic arable fields is generally due to the ban on synthetic N-fertilisers and herbicides. The limited availability and input of nitrogen, the application of mechanical and thermal weed control and more diverse crop rotations and a higher crop diversity lead to more favourable conditions for many wild plant species.

In organic grassland, the average number of species was found to be around 25 percent higher in organic than in conventional grassland, including some species in decline. The plant community structure in organic grassland is more even and more typical for a specific site than in conventional agriculture. The higher floral diversity is often caused by lower stocking rates and lower fertilisation levels in organic farms as well as to later mowing dates, especially in meadows, which means that grass species can reach the flowering stage and thus have a higher reproduction rate.

The limited availability and input of nitrogen, the application of mechanical and thermal weed control and more diverse crop rotations and a higher crop diversity lead to more favourable conditions for many wild plant species.

Vegetation on Faunal Diversity

Weeds influence the diversity and abundance of arthropods (e.g. beetles, ants and spiders), acting as natural food resource and shelter. Weeds like Umbelliferae, Leguminosae and Compositae play an especially important ecological role as they provide food and thus improve reproduction of many arthropod species. Research carried out on tomato plots on the effects of weed control on surface-dwelling arthropod species found the abundance of species is clearly influenced by weed biomass. Species numbers were lowest where mulching with rye straw was controlling the weeds. However, removing weeds within 20 cm of each plant reduced weed biomass but retained higher arthropod population than in the plots treated with herbicide or mulch.

With regard to pollinators, which greatly benefit from a richness of flowers, the fact that flowering weeds are more diverse and more abundant in organic arable fields and in organic grassland compared to conventional fields, where only few species and numbers were found, is particularly important. Many insect species which feed on nectar and pollen have a higher reproduction rate when they benefit from a better food supply through species-rich plant communities.

It may be assumed that, as for butterflies, the organic system also favours an abundance of further pollinators like bees and wasps. Flowering plants are also important for many beneficial arthropods such as predators and parasitoids. Orchards with rich floral undergrowth have a lower incidence of insect pests than orchards treated with herbicides, mainly because of an increased abundance and efficiency of predators and parasitoids. Orchards with rich floral undergrowth have a lower incidence of insect pests than orchards treated with herbicides, mainly because of an increased abundance and efficiency of predators and parasitoids.

In order to diversify the farming system and attract beneficial arthropods and pollinators, wild flower strips are sown in organic agriculture orchards. In a Swiss organic orchard, it was found that the strip management favoured beneficial insects and spiders, which reduced the density of aphids. The density of aphids was reduced due to higher mortality caused by increased numbers of predators feeding on aphids. Measures aimed at managing appropriate habitats and thus increasing floral and structural diversity is a key strategy for improved natural pest control.

Faunal Diversity

Organic agriculture displays in most cases a higher faunal biodiversity than conventional agriculture. Apart from the better food resources in organic fields, the key factors are more fauna-compatible plant protection management, organic fertilisation, the more diversified crop rotation and the more structured landscapes with semi-natural habitats and field margins.

The effects of organic agriculture on faunal biodiversity have been studied particularly for soil fauna and for birds. A review of 44 research studies in Europe and the United States of America (on farm and pilot trials) on the effects of farming systems on beneficial invertebrates and birds consistently shows a better performance of the organic system. Of the faunal groups analysed (i.e. earthworms, arthropods and birds) in 49 out of 55 investigations organic agriculture performed better in terms of abundance. As regards species diversity, in 15 out of 23 investigations organic agriculture performed better; in no case conventional farming had a better performance.

Earthworms

Earthworms are highly suitable bio-indicators of soil fertility, and they are known for their sensitivity to synthetic pesticides and to many agricultural practices. Due to their biology, earthworm population can indicate the structural, microclimatic, nutritive and toxic situation in soils. In conventional agriculture, earthworms are affected by the use of harmful pesticides and intensive soil cultivation.

Earthworms generally increase nutrient cycling rates. Their casts greatly help to improve soil structure and have high concentration of nutrients in an accessible form to plants. The burrowing activity of earthworms enhances aeration, porosity and drainage of the soil, all of which are important factors in the development of healthy and well-developed crop root system. Earthworms also play an important role in pest and disease control, including the reduction of leaf miner pupae and scab pathogens in orchards. Many investigations in Europe and North America provide evidence that generally, organically managed soils exhibit a higher abundance and species number than conventionally-managed plots or farms.

The biomass of earthworms in the organic system in the Swiss DOC long-term trial was 30 to 40 percent higher than in the conventional system, the number of individuals even 50 to 80 percent higher. Vertically-burrowing earthworm species which are of high agro-ecological relevance (e.g. for water infiltration, soil aeration, reduction of erosion) were enhanced by organic agriculture. Reasons for the higher presence of these species in the biological plots might be species-specific effects of pesticides. Vertically burrowing species are more exposed to pesticides because they are crawling on the ground surface for feeding and mating.

Organic matter constitutes an important food source for earthworms and can be maintained by an appropriate fertilising and crop rotation system. Investigations have shown that earthworms also benefit greatly from green manuring and the planting of grass-clover in the crop rotation. The less intensive farming practices (plant protection, soil cultivation), the more favourable fertilising management and the more diversified crop rotations on organic farms enhance specimen- and species-rich earthworm population.

Arthropods

Beneficial arthropods that live above the ground include ground beetles (carabids), rove beetles (staphilinids) and spiders. Many of these polyphagous species, which feed on a wide range of food items, are important predators, and in arable crops they play an important role in the regulation of various pests. In addition, certain arthropods, especially carabids, are considered as sensitive indicators of habitat quality. Monitoring them yields useful information on the sustainability of different agricultural farming systems. In conventional agriculture, synthetic pesticides can have negative impacts on beneficial arthropods. Pesticides affect arthropods either directly, via contamination or reduction of their prey, or through alterations of the micro-habitat. The reproduction rate may be reduced by sublethal long-term effects of synthetic pesticides used in conventional agriculture. Foliar fungicides can lead to mortality in springtails and thus negatively influence polyphagous predators.

Higher fertilisation in conventional fields leads to higher crop density, and this in turn can alter the micro-climate as well as reducing the occurrence of species dependent on a warm micro-climate and on light. This phenomenon was found in the DOC long-term plot-trial as well as on farm level. Organic fertilisation can enhance epigeic arthropods through a richer supply of mesofauna decomposing organic compounds. In several investigations of on-farm sites as well as in plot trials, a higher diversity and abundance of arthropods was found in organic and biodynamic plots compared to conventional plots. In the DOC long-term trial, the organic and biodynamic treatment had up to 100 percent more carabids, 60 to 70 percent more staphylinids and 70 to 120 percent more spiders. The organic fields were also characterised by a generally more even distribution of species, which means that the community is less dominated by few species only.

In several investigations of on-farm sites as well as in plot trials, a higher diversity and abundance of arthropods was found in organic and biodynamic plots compared to conventional plots. Researchers have also found greater diversity and abundance of many other invertebrate species in organic agriculture systems. For example, a study comparing spider communities in organic and conventional winter wheat fields in the United Kingdom found the abundance and diversity of spiders greater on the organic fields. It was concluded that the results were highly affected by the increased levels of under storey vegetation (i.e. broad-leaved and grass species) in the organic fields. Significantly more non-pest butterflies were recorded on organic farms than conventional farms but there was no significant difference in the abundance of pest species. The quality and amount of food are key factors for the survival of the arthropod population, which find a higher proportion of suitable food resources in organic fields including flowering non-grass plants.

Birds

Birds are well-suited indicator organisms showing the environmental status of nature and landscape infrastructure including agricultural land. Many bird species feed on insects, and an abundant presence may thus also contribute to a better natural pest control. The breeding population of a number of bird species in Western European farmland have declined in the past decades. This development is probably linked to intensified agricultural practices, including a general reduction in crop diversity, an increase in average field size, simplified rotations, cultivation of natural habitats, drainage, and increasing use of fertilisers and synthetic pesticides. Investigations have found a relationship between pesticide use and a decrease in quality and amount of food available for birds on conventionally farmed land. A consequence is a decreased breeding success of birds.

Several studies show that bird densities are higher in organic farms. A study by Rhone-Poulenc has shown a steady annual increase in the number of bird territories on land converted to organic production and a higher overall number of territories on the organically managed land. A study by the British Trust for Ornithology compared breeding and over-wintering of birds on 44 organic and conventional farms. The study concluded that breeding densities of skylarks were significantly higher on organic farms and generally higher densities of birds, especially in winter, were found on the organic farms.

A three-year study in Denmark concentrated on the non-crop habitats, such as hedgerows, of conventional and organic farms and their effects on bird population. The abundance of birds was 2 - 2.7 times greater on the organic farms. In total, 24 species where more prevalent on organic as opposed to conventional farms, of these 11 species had declined in number in Denmark since 1976. Christensen et al. investigated a total of 31 farms comparing the bird densities on organic land with those on conventionally managed land. On one of the sites investigated they found a mean population density of 22.8 pairs of breeding birds per ten hectares on the organic land compared to 9.9 pairs on the conventional land. Also flora and invertebrate fauna in organically versus conventionally managed land had been compared on the farms selected for the bird survey.

By late June, the organic fields had 50 to 70 percent more species of wild plants than the conventional fields, with significantly more weeds and biomass. These differences increased markedly during the growth season due to herbicide treatment in the conventional areas. The average biomass in the organic fields was significantly more stable than in conventional fields, resulting in greater security of food supply. These results suggest that limited food availability is a key factor in the reduced number of birds in conventionally farmed areas.

In the coffee producing countries of Latin America, growing organic coffee and cocoa under shade can have a major impact on biodiversity. Research carried out by the Smithsonian Migratory Bird Centre in Colombia and Mexico showed over 90 percent fewer bird species in sun-grown coffee plantations as opposed to shade-grown coffee. Although organic standards do not explicitly state the need for coffee to be grown in the shade, shade-grown practice is recommended as it fulfils requirements to enhance soil fertility, pest and disease control and expands crop production option.

Little work is available on the effects of organic agriculture on other wild animal groups. This could, however, be an important research theme, particularly with respect to pollinators. Many plants require pollen from other individuals to seed and regenerate. Reasons for the higher number and species diversity of birds and bird territories on organic farms include better breeding habitats and better food conditions. Over 80 percent of all flowering plant species and over three-quarters of the major world crops rely on animal pollinators. The principle pollinators are bees: approximately 73 percent of the world's cultivated crops, such as cashews, squash, mangoes, cocoa, cranberries and blueberries, are pollinated by some variety of bees, 19 percent of flies, 6.5 percent by bats, 5 percent by wasps, 5 percent by beetles, 4 percent by birds, and 4 percent by butterflies and moths.

Of the hundred principal crops that make up most of the world's food supply, only 15 percent are pollinated by domestic bees (mostly honey bees, bumble bees and alfalfa leafcutter bees), while at least 80 percent are pollinated by wild bees and other wildlife (as there are an estimated 25 000 bee species, the total number of pollinators probably exceeds 40 000 species). Services of native pollinators are estimated to be worth US$4.1 billion a year to United States agriculture alone. Pollinators provide an essential ecosystem service that contributes to the maintenance of biodiversity, and ensures the survival of plant species including plants that provide food security to numerous households.

Declines in population of pollinators now threaten both the yields of major food crops and the survival of wild plant species. Due to an epidemic of mites, a quarter of North America's wild and domestic honeybees have disappeared since 1988, with a cost to American farmers of US$5.7 billion per year. Currently, 82 species of mammalian pollinators, including bats, 103 species of avian pollinators and one reptile are considered threatened or extinct according to IUCN criteria, the ratio of threatened vertebrate pollinators to the total numbers of vertebrates in their genera being extremely high. Geographically, large numbers of vertebrate pollinators are at risk in Australia, Colombia, Ecuador, Indonesia, Madagascar, Mexico, Papua New Guinea, Peru, and the United States.

Common threats include: loss of nesting and roosting sites, habitat fragmentation by conversion or destruction of vegetation, habitat fragmentation by excessive exposure of nectar plants to herbicides and pollinators to pesticides, over-hunting, disruption of nectar corridors required by migratory pollinators, and competition by invasive species.

By not using synthetic chemical pesticides or herbicides and enhancing ecosystem diversity, it may be assumed that organic agriculture benefits pollinators and contributes to their conservation and survival.

According to the World Resources Institute, an ecosystem is made up of the organisms of a particular habitat, such as a farm or forest, together with the physical landscape in which they live. Although little research has been carried out comparing agro-ecosystem diversity in different farming regimes, many of the principles of organic agriculture are likely to have a positive impact on ecosystem diversity.

Ecosystem Diversity in The Standards for Organic Agriculture

The 2002 IFOAM Basic Standards for Organic Agriculture Production and Processing include principles and recommendations on "organic ecosystems" where provisions are made to "maintain a significant portion of farms to facilitate biodiversity and nature conservation", including (among others) wildlife refuge habitats and wildlife corridors that provide linkages and connectivity to native habitat.

The Swiss organic standards require organic farmers to use 7 percent of their land as semi-natural habitats, including field margins. Many other organic certification bodies have included biodiversity requirements in their standards. This applies to organic and low-input farmers receiving direct payments under the agri-environment schemes. Direct payments by the state are linked to ecological performance which is controlled by inspection services.

Organic agriculture has a high and possibly decisive potential for reversing the dramatic decline of biodiversity. In Sweden, a working group made up of organic farmer groups, nature conservationists, Government agencies and universities has been working since 1997 to strengthen the links between organic agriculture and biodiversity conservation, and they set up biodiversity plans for organic farms. The main objectives, apart from helping organic agriculture enhance biodiversity, have been to start cooperation and dialogue between the nature conservation and organic agricultural movements and to spread knowledge about biodiversity in organic agriculture. From there, the discussions developed into a planning exercise leading to proposals for changes to the country's main organic standard (set by the certification body, KRAV), to require all organic farmers to have a plan for the management of biodiversity on their farms from 2001.

Organic agriculture has a high and possibly decisive potential for reversing the dramatic decline of biodiversity. However, further efforts must be taken to refine the basic organic standards for biodiversity and landscape. Actual experiences show that there is a considerable and diverse synergy potential between biodiversity conservation and farm income. Biodiversity issues based on measurable performance are more and more important as a basis for subsidies and, at least in some countries, for consumer decisions for buying organic products.

Semi-Natural Habitats

The conservation and the management of semi-natural habitats play an important role in organic agriculture. Semi-natural habitats are refuges for endangered plant species which in former times were found in meadows and arable fields. Semi-natural habitats and field margins are also essential for the survival of many invertebrates, especially due to favourable food and over wintering conditions. They also function as habitat cross-links between meadows, fallows and different field margins.

Quantitative data from Switzerland found that the proportion of semi-natural habitats and field margins per farm to be 16 percent on organic farms, compared to 3.7

percent on conventional farms. It was also noted that in Germany organic farmers took specific measures to increase the habitat diversity (hedges, low-input orchards, ponds, corridors, habitat networks, wildlife refuges and devices) as well as to better connect farmed areas and the surrounding habitats.

A recent study, also in Switzerland, shows how important the combination and integration of semi-natural habitats with organic farms is for the conservation and enhancement of species diversity and abundance of beneficial arthropods. In this study organic agriculture had been compared with low-input integrated crop management within the framework of a paired farm survey. Also nearby semi-natural landscapes have been used to compare the effects of organic agriculture and low-input farming system on carabid beetle and spider fauna. Many endangered or rare species which are enhanced by semi-natural habitats and field margins were more abundant in organic arable fields than in the integrated managed fields.

Also, agro-ecologically important carabid species and wolfspiders were found in higher numbers on the organic farms. This aspect indicates that the improvement of landscape infrastructure in combination with organic agriculture may be an important factor for the conservation and enhancement of species-rich communities on agricultural land.

Landscape Development

Organic agriculture has the potential for a positive landscape development because it itself need a rich and diverse landscape infrastructure with semi-natural habitats. By studying the whole farm, or even the whole landscape within which the farm operates, researchers are trying to find ways to characterise the benefits of organic agriculture to the ecosystem. A research project funded by the European Union found that the diversity of landscape and farming systems was greater in organic farms, regarding land use types, crops, livestock, plantings (hedges, solitary shrubs, trees) and flora. In terms of landscape diversity, the organic types of agriculture have a good potential for positive contributions to a sustainable agrarian landscape.

Protected Areas

Protected areas, as defined by the World Commission on Protected Areas, are areas especially dedicated to the protection and maintenance of biological diversity. Protected areas are not always strict nature reserves. Instead, they can fulfil many functions alongside biodiversity conservation. Because of its biodiversity benefits, organic agriculture offers an important agricultural management option in several protected area categories.

Because of its biodiversity benefits, organic agriculture offers an important agricultural management option in several protected area categories. In Italy for instance, the Italian Association for Organic Agriculture (AIAB) project on "Organic Agriculture and Agro-ecology in Regional Parks" has been working with the regional park authorities in the Emilia-Romagna region to promote organic agriculture in relation to the regional agri-environment programme. During the first two years of activity there was an increase in the rate of adoption of the regional agri-environment programme, particularly of organic agriculture, by farmers in the park and the buffer zone. Between 1996 and 1997, 113 farms in the area applied for organic certification, compared with only 73 between 1994 and 1996.

Buffer Zones

A secondary and closely connected link between ecosystem

diversity, protected area management and organic agriculture is in the buffer zones (the region near the border of a protected area). Buffer zones are by their nature areas where land management aims to help maintain the integrity of the ecosystem of the core protected area. Where agriculture is the dominant land-use in buffer zones, the detrimental effects of farming systems can be reduced by conversion to organic systems. The use of organic agriculture in protected buffer zones has been explored in the Meso-American Biological Corridor, a complex of protected areas and sustainable management stretching over seven countries. The initiative envisages a range of sustainable land uses within the buffer zones and linking areas, including certified forest management and organic agriculture.

In Peru, organic agriculture practices are promoted in the buffer zone of the Ampay Forest Sanctuary. Thanks to information measures, the awareness of the population about the biological importance of the sanctuary was increased, and as a result the area has converted into a high-priority conservation area for ecotourism, governmental and private investment in infrastructure, promotion and conservation of natural resources. Where agriculture is the dominant land-use in buffer zones, the detrimental effects of farming systems can be reduced by conversion to organic systems.

There is increasing recognition among nature conservation organisations that many species interact with agricultural systems, even if their primary habitat is in natural areas. The management of these agricultural systems can, thus, dramatically affect overall levels of biodiversity, as well as the success of particular species. The German conservation organisation BUND claims that traditional concepts of nature protection have failed because conservation goals have not been achieved, even in protected areas, due to agricultural intensification in the surrounding areas. Many conservation organisations are therefore calling for a general extensification of agricultural land use and promote organic agriculture for that reason.

In 1999, a joint workshop in Vignola, Italy, organised by The World Conservation Union (IUCN) and IFOAM together with the World Wide Fund for Nature (WWF), was held to exchange ideas and information on organic agriculture and biodiversity. A joint action plan for both the nature conservation and organic movements was drawn up in the so-called Vignola Declaration.

In both developed and developing countries, one of the driving forces behind the growth of organic agriculture is the nature conservation agenda. Though less significant than market prospects and development prospects, a recent literature survey highlights a number of examples where nature conservation organisations are working closely with local farmers who live in, or close to, areas of significant nature conservation interest. In the last twenty years, governments also have been gradually moving towards policies aiming at encouraging the links between organic agriculture and biodiversity conservation. In the European Union, environmental concerns have gained a more prominent place in agricultural policy. The most notable contribution of the Common Agricultural Policy of the European Union towards more environmentally sustainable systems has been the introduction of the agri-environment programmes in 1993, which now continue under Agenda 2000. The impetus for some of these initiatives comes from the Convention on Biological Diversity.

The implementation of agri-environmental measures over the whole of the European Union and also other European countries is the core of the European Community's environmental strategy. So far, organic agriculture has played a central role in most countries' national agri-environment policy. The main reason for this policy support has been the perceived positive environmental effects of organic agriculture.

Also in other parts of the world, governments and development agencies are beginning to promote organic agriculture, especially in protected areas, because of its biodiversity benefits. Organic agriculture brings several environmental benefits but the question arises on the threshold for converting land to organic management. One approach to determining this is to examine the benefits obtained from moving away from conventional agriculture.

One estimate of the negative environmental externalities from conventional agriculture in the United States is derived from Pimentel et al. who estimated total indirect private and external (off-farm) costs of the use of synthetic pesticides at US$12.1 billion annually: this included the private input costs of US$4 billion and US$5 billion in environmental and health costs. Farmers had a US$12 billion incentive to use these pesticides. By comparison, the private benefits captured by the farmers were calculated at US$16 billion but in their decision they were ignoring the costs of US$8.1 billion imposed on the rest of society. These rather dated estimates indicate the nature and magnitude of the problem.

Redman estimated the costs of cleaning drinking water in the United Kingdom in excess of US$2 billion in initial investments plus US$240 million annually. These costs relate to pesticides and nitrate pollution. So far, organic agriculture has played a central role in most countries' national agri-environment policy. Another approach to estimating costs is to look at the costs of compliance to environmental regulations. A study comparing the impacts of environmental standards on agriculture in Australia, Canada, the European Union, New Zealand and the United States concludes that the compliance costs would normally be less than 3 or 4 percent of gross revenue, although intensive livestock industries face the greatest external costs.

Agricultural-related environmental problems relate chiefly to water quality and intensive livestock operations. To illustrate this point, research using the United Kingdom Government data suggests that the costs of meeting the Nitrates Directive in England and Wales reduced net farm income by 2.4 percent in the poultry sector, 11.3 percent in the pig industry, 4.5 for beef and 9.6 on dairy farms. Most of the cost relates to capital expenditure to adequately handle manure. Animal welfare regulations for poultry and pigs are estimated to reduce net farm income by 38 and 6 percent respectively.

Crops do not face the same compliance costs. The implications of these estimates are that perhaps environmental concerns do not justify the abandonment of pesticides and synthetic fertilisers that a widespread shift to organic agriculture would entail. However, this approach ignores the environmental damage that occurs even though the regulations are being met. There is some evidence that consumers who have recently entered the organic market are less motivated by ecological issues and are more interested by organic food as a safe product. What is interesting is that sales of organic products have increased markedly as a result of recent food scares, suggesting that new consumers are concerned about the product itself as well as, or perhaps rather than, the environment in which it is produced. However, the increases in sales have not related to the particular products (that is, beef) to which the scares applied.

ORGANIC AGRICULTURE AND CLIMATE CHANGE

Global climate change is considered one of the most urgent environmental problems. The main negative impact on climate change is the emission of greenhouse gases (CO_2, CH_4, N_2O), which is directly or indirectly due to the burning of non-renewable resources (carbon bound in mineral oil or coal). Tropical rainforests hold the biggest living biomass on very delicate soils that may lose their fertility completely when clear cutting is performed as in recent decades. Agriculture contributes to over 20 percent of global anthropogenic greenhouse gas emissions. Moreover, agricultural intensification has had major detrimental impacts on the terrestrial and aquatic ecosystems of the world. The doubling of production during the last 35 years was associated with a 6.9 fold increase in nitrogen fertilisation, 3.5 fold increase in phosphorus fertilisation and a 1.7 fold increase in irrigated land.

Agriculture, however, is not only contributing to global warming, but is also affected by it to a major extent. According to Burdick increasing global warming will shift cultivation zones polewards, plant growth and production being jeopardized by changes in the distribution of rainfall, the increase of UV-B radiation, and changes in the chemical composition of the atmosphere. In regions with continental climate, soils are subject to desiccation, meaning climate change will aggravate problems of salinity, erosion, and desertification. Extreme climatic events will occur more frequently. Pests and diseases favoured by a warmer climate will continue to proliferate. All these factors will have negative impacts on agricultural yields.

Since climate change has a direct impact on agriculture, environmentally sound farming methods need to be developed and practised. Organic agriculture not only enables ecosystems to better adjust to the effects of climate change but also offers a major potential to reduce the emissions of agricultural greenhouse gases. Moreover, mixed farming and the diversity of organic crop rotations are protecting the fragile soil surface and may even counteract climate change by restoring the organic matter content. The carbon sink idea of the Kyoto Protocol may therefore partly be accomplished efficiently by organic agriculture.

Carbon Dioxide

Cultivation of land with the clear cutting of natural forests has been the main CO_2 emission that may be ascribed to the agricultural sector. Soil processes with respect to carbon are characterised by the dynamic equilibrium of input (photosynthesis) and output (respiration). Principally, all the organic matter entering the soil is mineralised. Changing environmental conditions and land management may induce a temporary or definite change of the equilibrium to a new level that is considered stable.

Emissions

Conversion of natural vegetation to agriculture is a major source of CO_2, not only due to losses of plant biomass but also, increased decomposition of soil organic matter caused by disturbance and energy costs of various agricultural practices such as fertilisation and irrigation. CO_2 emissions from the agricultural sector represent 21-25 percent of total CO_2 emissions, due to fossil fuels used on farms, shifting patterns of cultivation and chiefly, deforestation.

Appropriate management practices could increase carbon sinks and energy efficiency improvements and production of energy from crops and residues would result in a further mitigation potential, or a cumulative carbon storage.

In order to compare farming systems, they are subdivided into emissions due to burning of oil and fuel (direct energy) and the use of oil and fuel for the production and transport of fertilisers, machinery and synthetic pesticides (indirect energy). CO_2 emissions per hectare of organic agriculture systems are 48 to 66 percent lower than in conventional systems. Haas and Kopke calculated the

CO_2 emissions of German organic farms to be 0.5 tonnes of CO_2 per hectare whereas in conventional agriculture the amount was 1,3 tons, a difference of 60 percent. The main effects of organic agriculture that are responsible for this difference are:

— the maintenance and increase of soil fertility by the use of farmyard manure;

— the omission of synthetic fertilisers and synthetic pesticides;

— the lower use of high energy consuming feedstuff.

In organic agriculture, almost 70 percent of the CO_2 was due to fuel consumption and the production of machinery, while in conventional systems 75 percent of the CO_2 emissions are ascribed to N-fertilisers, feedstuff and fuels.

Soils as a Sink for Atmospheric CO_2

Soil carbon levels have decreased under agricultural land use. Sustainable agricultural strategies comprising recycling of organic matter, tightening internal nutrient cycles, and low- or no-tillage practices may rebuild organic matter levels and reduce losses from the system. Mixed farming with manure amendment leads to higher organic matter levels in soil.

Combined with other organic agriculture techniques Drinkwater et al showed, in a long-term experiment, a considerable gain in soil organic matter compared to a conventional system at comparable yield. Many long-term experiments in the world support the cognition that organic fertilisation (animal manure, green manure, catch and cover crops) rebuild soil organic matter. The accumulation of organic matter in the soil is restricted by soil type, climate and management factors and will reach a certain saturation level.

Haas and Kopke calculated that, despite generally lower crop yields, plant productivity in organic agriculture accounts for almost the same organic matter return as in a conventional system. This is due to the more diverse crop rotations and the longer period the soil is covered by plants. On the other hand, organic crop yields in developing countries may be considerably higher than the national average, which has implications on organic matter return and carbon sinks.

A 20 percent increase in soil organic matter as a result of organic agriculture would result in an estimated amount of 9 tonnes carbon per ha, which is in line with the results over a 15-year period. Smith et al calculated a considerable potential increase of soil carbon when manure, straw-recycling, minimal tillage, reforestation and energy-saving plant production are combined.

Nitrous Dioxide

Nitrous oxide emissions not only contribute severely to the greenhouse effect but also to the depletion of stratospheric ozone. Almost 90 percent of the global atmospheric N_2O is formed during the microbial transformation of nitrate (NO_3-) and ammonia (NH_4^+) in soils and water.

Globally, agriculture contributes to 65-80 percent of total N_2O, mainly from nitrogenous fertilisers on cultivated soils, cattle and feedlots. In OECD countries the agricultural contribution to N_2O emissions is estimated at 58 percent. N_2O emissions from soils are due to the unproductive loss of mobile N.

Any nitrogen input (mineral and organic fertilisers, biologically fixed N, crop residues) and also the mineralisation of nitrogen compounds in soils rich in organic matter contribute to the emission of N_2O. Especially in agricultural soils elevated N_2O production depends on the nitrogen fertilisation level. The above-mentioned inputs are proposed as determinants of N_2O emissions by the Inter-Governmental Panel on Climate Change.

Tropical ecosystems, currently an important source of N_2O (and NO) are often phosporus-limited rather than being N-limited like the Northern hemispheric terrestrial ecosystems. Nitrogen fertiliser inputs into these phosphorus-limited ecosystems generate N_2O (and NO) fluxes that are 10 to 100 times greater than the same fertiliser addition to nearby N-limited ecosystems.

The ban on synthetically-produced mineral nitrogen in organic agriculture confines the productivity to the natural system limits (i.e. N-fixation) or the limits defined by the annual nutrient balance of the farm, including imported fodder and organic fertilisers. Organic agriculture therefore is likely to emit less N_2O due to restricted intensity because of:

— a systemically lower N-input;

— less N from organic manure due to lower livestock densities;

— a higher C/N ratio of applied organic manure and less available mineral nitrogen in the soil as a source for denitrification;

— the permanent plant cover in organic systems which results in a more efficient uptake of mobile nitrogen in soils, thus reducing the potential and risk for N_2O emissions.

As Tilman points out "sustainable and productive ecosystems have tight internal cycling of nutrients, a lesson that agriculture must relearn".

Methane

Agriculture is believed to account for roughly two-thirds of the total man-made CH_4 mainly from paddy rice fields, burning of biomass and ruminants (enteric fermentation and animal waste treatment). Aerobic agricultural soils, however, are considered sinks for atmospheric CH_4. Fertilisation with mineral nitrogen has been shown to inhibit CH_4-oxidation in soils. The lower N-fertilisation level in organic agriculture may therefore exhibit an advantage for CH_4-oxidation. However, no data exist on the effects of organic agriculture on methane emissions.

The emission of methane by ruminants is probably not affected by organic production. The higher proportion and lower productivity of ruminants in organic agriculture may, however, lead to slightly higher emissions of CH_4.

On the other hand, standards and breeding programmes aim at longevity in order to prolong the productive period in relation to the unproductive life of young cattle. Correspondingly the "unproductive" CH_4-emission of calves and heifers may be reduced. Organic agriculture not only enables ecosystems to better adjust to the effects of climate change but also offers a major potential to reduce the emissions of agricultural greenhouse gases.

Desertification

The degradation of drylands is called desertification. Desertification is caused by overcultivation, overgrasing and deforestation and results in soil exhaustion and soil erosion. It diminishes soil productivity, reduces food production, deprives the land of its vegetative cover, and negatively impacts areas not directly affected by its symptoms, by causing floods, soil salinisation, deterioration of water quality, and silting of rivers, streams and reservoirs.

Organically-managed soils have a high potential to counter soil degradation as they are more resilient both to water stress and to nutrient loss. Organically-managed soils have a high potential to counter soil degradation as they are more resilient both to water stress and to nutrient loss. Organic farmers feed their soils with organic fertilisers, and they can thus enhance degraded and problematic soils. The water and nutrient retention capacity is increased thanks to a high level of organic matter and permanent soil cover. Microorganisms have a good feeding base and create a stable soil structure. Due to the resulting high moisture retention capacity the amount of water needed for irrigation can be reduced substantially.

There is little scientific evidence demonstrating organic agriculture's potential for combating desertification but several practical examples of organic agriculture systems in arid areas show how organic agriculture can help bring degraded lands back to fertility. A SEKEM biodynamic farm in Egypt cultivates 70 hectares of desert near Cairo. By using organic and biodynamic agricultural methods (composting, mulching, cover cropping) the desert sands were converted into fertile soil, supporting livestock and bees.

Organic agriculture can create suitable micro-climates in dry areas. For example in Kenya, the International Centre for Research in Agroforestry runs organic agriculture projects to fight drought. Agroforestry generates multiple benefits, including erosion control and moisture retention. In Tanzania, the Chagga home gardens on the slopes of Mount Kilimanjaro, where certified organic coffee is produced, provide an excellent example of agroforestry.

The system includes a diversity of cash and subsistence crops (e.g. bananas, coffee, yams, and beans) as well as livestock. Cattle and pigs are kept in stables ("zero grazing") and manure is recycled as fertiliser. The home gardens are designed to maximise diversity. Elaborate patterns of vertical zonation provide a range of sunny and cooler conditions for different species.

Organic agriculture can counter erosion successfully. While in conventional agriculture in the tropics even flat soil gets eroded due to the use of herbicides and the lack of soil cover, in organic agriculture permanent soil cover is an intrinsic part the system. In trials on a Cuban citrus plantation, the Cuban Citrus Institute and Swiss FiBL use Teranamus labialis, Arachis pintoi, Neonotonia wightii and other legumes.

These locally-adapted leguminous crops can help restore degraded soils very fast, successfully suppress weed, fix nitrogen and prevent erosion. FiBL-experiments have also shown that cover crops do not compete with the main crops for nutrients or water. The precondition is, however, that the system be properly set up and adapted to soil and climatic conditions. Soil life is enormously enhanced, and water retention capacity is increased. The cover crop plants also serve as habitats for predators.

Organic agriculture can also counter salinisation problems. The Fazenda Tamanduà in Brazil is situated in an area severely affected by salinisation caused by inappropriate irrigation techniques. It is a certified organic farm with more than 3000 hectares of which 650 hectares are cropped with mango trees. Due to the arid conditions, irrigation is the prerequisite to agricultural production. Conventional agriculture in the area uses the water from the rivers and causes further salinisation by over-irrigation with already salinated water.

Fazenda Tamanduà limits its water use to rain water, which is collected in rainwater collection basins, thus avoiding the depletion of the water resources. At the same time the salinisation of soils is considerably reduced. Fertilisation regimes include grazing of cattle under mango trees as well as application of composted manure. The organic substance acts like a sponge in the soil, and fields thus fertilised have a higher moisture retention capacity and therefore need less irrigation.

Even though the adoption of organic agriculture seems a feasible alternative in arid areas, there is a range of constraints to adopting organic agriculture systems or even individual techniques. These include a lack of knowledge, lack of organic materials, insecure land tenure, the perception of organic agriculture as being old fashioned and the fact that extension services promote chemical farming. The promotion of organic agriculture through education and training, in countries suffering desertification, could be a key to bringing degraded land back into production.

REFERENCES

Alger, K., *The reproduction of the cocoa industry and biodiversity in southern Bahia, Brazil,* paper presented at Cacao Workshop in Panama, Smithsonian Migratory Bird Centre, 1998, available at www.nationalzoo.si.edu/smbc/Research/Cacao/alger.htm.

Alonso Villalon, N., "First Spanish aquaculture certification includes sturgeon", *The Organic Standard* 12: 14-15, 2002.

Barbera, G. & La Mantia,T., *Analisi agronomica energetica, Filiere atte allo sviluppo di aree collinari e montane: il caso dell'agricoltura biologica,* Chironi G Vo. 1 RAISA-University of Palermo, 1995.

Barret, J.R., "Livestock farming: eating up the environment?", *Environmental Health Perspectives,* 109(7): A312-A317, 2001.

Cederberg, C. & Mattsson, B., "Life cycle assessment of Swedish milk production: a comparison of conventional and organic farming", In D. Ceuterick, ed. *Proceedings of the international conference on life cycle assessment in agriculture, agro-industry and forestry,* 3-4 December, Brussels, Belgium, 1998.

Christensen, D.K., Jacobsen, E.M. & Nohr, H., "A comparative study of bird faunas in conventionally and organically farmed areas", *Dansk. Orn. Foren. Tidsskr* 90: 21-28, 1996.

8

Certified Organic Agriculture

Organic agriculture has a number of distinct characteristics. These features determine the driving forces behind the growth in organic production and sales. They are:

— organic agriculture is a process-oriented, rather than a product-oriented, production system;

— the organic agriculture process involves significant constraints that raise costs of production and marketing;

— consumers buy organic products mainly for the perceived health, food safety and environmental benefits.

PROCESS-ORIENTED PRODUCTION

The most notable characteristic of organic agriculture is the emphasis on the production process rather than a product itself. With a credible certification scheme, the consumer is ensured that products have been produced according to certain defined standards and are likely to be:

— low in pesticide residues;

— grown in a way that is favourable to the environment and possess other characteristics such as the humane treatment of animals;

— provide fair payments to producers, particularly in developing countries.

PROCESS-ORIENTED, RATHER THAN A PRODUCT-ORIENTED, PRODUCTION SYSTEM

These characteristics cannot be seen in the product itself. Consumers may therefore not have a good understanding of what they are buying, and may be more inclined to switch to non-organic products offering somewhat similar characteristics, such as "free range" rather than organic eggs. Future organic sales may compete with products carrying ecolabels, possibly without verification and at a reduced cost as compared to organic products.

Production and Marketing Costs

The primary feature of all organic production methods is the restriction on the use of:

— synthetic fertilisers and pesticides in crop and fodder production;

— synthetic health care products, growth promoters and hormones in livestock production;

— synthetic preservatives and irradiation in post-harvest handling;

— GMOs at all stages in the food chain.

These constraints tend to require changes in the rotation away from monocultures and, in crop dominated systems, towards some livestock. Crop yields are generally lower on organic farms when compared with high input agricultural systems, especially during conversion years. For example, in Europe, cereal yields were commonly observed to be about 60-70 percent of those obtained under high external input management. Stocking rates on organic pasture also tend to be lower.

As livestock generally provides a lower return per hectare than crops, organic cereal/livestock producers are doubly disadvantaged as they have lower yields, especially during conversion, and a less favourable output mix. However, this may be compensated for by higher output prices as well as lower input costs, and many European organic farmers also receive additional taxpayer-funded subsidies.

Economic performance on organic farms in many countries and numerous crops are often found to be comparable with those on conventional farms, even without additional subsidies and sometimes even without output price premiums. For example, a multi-country study of European organic agriculture concludes that "the profits of organic farms are quite similar to those of comparable conventional farms." The authors duly note the usual caveats, in particular that the degree of variation within the samples is very high, highlighting the risks and opportunities available. In an earlier study, Lampkin and Padel mention similar results in a number of European countries. A Danish study using 1996 data found that all three types of organic farms (cereal, dairy and pig) were similarly or more profitable than their conventional counterparts. In general, it is easier to convert dairy farms profitably than intensive livestock operations.

The bulk of the cost to the consumer is post farm-gate costs for all food products, conventional and organic. A loaf of bread costs the consumer ten to twenty times what the producer receives for the wheat, partly due to the processing involved (storage, handling, distribution, milling, baking and retailing). But even relatively unprocessed products such as milk are three times as expensive at the retail level. Marketing costs for organic products in particular are higher because of the need for segregation to preserve organic identity. Also, certification and the scale of distribution contribute to high costs. As organic agriculture assumes larger shares of the market, economies of scale should be able to lower these costs significantly.

Consumer Benefits

It is not easy to generalise why consumers buy organic products. Many consumers believe that the products are better for people's health or have better organoleptic qualities. These claims have proven to be difficult to verify, although some recent reports suggest that there is a real difference in nutrient and vitamin status. For some people, food safety seems the major concern, whereas for others, environmental aspects appear paramount.

In comparison with conventional agriculture, organic production methods are considered to have less detrimental environmental effects. This is of course not surprising as the environment is the main objective of organic agriculture. Encouraging a move towards organic agriculture would reduce environmental clean-up costs and lower the damage associated with externalities. Subsidies to organic agriculture are touted as justified because of the environmental benefits they bring. Organic agriculture is one approach that can be used to achieve environmental quality targets. Policy-makers are starting to consider how best to facilitate organic agriculture development in order to maximise environmental benefits.

Organic Agriculture Gains Ground

Area under certified organic agriculture

The International Federation of Organic Agricultural Movements (IFOAM) collaborated with the Foundation for Ecology and Agriculture (SOL) in Germany to obtain up-to-date estimates of certified organic land. According to this study, the global area of land under certified organic agriculture in 2002 was 17.8 million hectares of which:

— 7.7 m ha are located in Oceania (mainly in Australia);
— 4.2 m ha in Europe (with Italy having the largest area, 1 million ha);
— 3.7 m ha in Latin America (with almost all of this in Argentina);
— 1.3 m ha in North America (USA nearly 0.9 million ha);
— 94 000 hectares in Asia;
— 60 000 in Africa.

The lack of a recognised system of organic agriculture data collection worldwide has inherent weaknesses in terms of reliable information. In fact:

— Argentina has 3.2 million ha of certified organic lands (instead of the 2.8 million ha reported by SOL);
— Uganda has 122 000 ha (instead of the 5 000 ha reported by SOL);
— South Africa has 45 000 ha (instead of the 6 000 ha reported by SOL);
— India estimates certified organic lands to 41 000 ha (while the SOL reports 2 800 ha).

Those few country examples alone suggest a global certified organic land area in excess of half a million hectares than publicly available figures.

Use of land

Less than half of the global 18 million hectares of certified organic land is dedicated to arable agriculture. In Australia, about 7 million ha of organic certified lands are extensive grazing lands. These lands belong to a marketing group (the Organic Beef Enterprises), which counts 30 members. One OBE member has the world's largest certified organic property (994 000 ha) as well as four other properties, totalling 3 248 290 ha. In Argentina, 91 percent of certified organic lands (or 2.9 million ha) are natural pastures for sheep and cattle (e.g. Patagonia) and forest areas for apiculture.

Also here, a single owner holds two extensive sheep grazing properties of 0.5 million ha. In both Australia and Argentina, the extensive nature of livestock systems is the most suitable management option in dryland conditions and large farms are typical. In contrast to these huge organic animal production properties, smallholdings characterise organic arable lands (with few exceptions). Less than half of the global 18 million hectares of certified organic land is dedicated to arable agriculture.

In the United States of America, at the end of the 1990s, well under 1 percent of crop and pastureland was certified as organic (i.e. 545000 ha) despite a rapid expansion (23 percent per annum) of retail sales during that decade. The US Department of Agriculture estimates suggest that the area of certified organic cropland in 1997 was 345 000 hectares, more than double the 1992 estimate. The most significant crops are wheat (50 000 ha), corn (17 000 ha), barley (12 000 ha) and oats (12 000 ha). Overall, organic grains showed a 23 percent annual growth in the two years since the previous survey in 1995. However, these crops comprise only 0.2 percent of the total United States cropped area in grains. In terms of inroads into conventional production, the most significant crops are spelt (37 percent of the conventional area), buckwheat (30 percent), herbs (37 percent) and mixed vegetables on plots of less than 2 ha (37 percent).

The area of organic pasture and rangeland in the United States amounted to 200 000 hectares in 1997, on par with 1992 but an 80 percent increase on 1995. Around one third of the certified organic area in 1997 was in Alaska and almost half was in Colorado, where productivity may not be as high as in other regions. There is little relationship between livestock numbers and certified organic area. Beef cattle numbers fell from 6 796 in 1992 to 4 429 in 1997. The large increases since 1992 have been in dairy cows (up from 2 265 to 12 897) and layer hens (44 000 to 538 000). Adding crop, pasture and rangeland together gives a total area under certified organic production, in 1997, of around 0.2 percent of United States agricultural land.

Data on certified organic agricultural land use in the European Union indicates an average annual increase of 24.5 percent in the five years to the end of 2000. The data revealed considerable variation across countries in the European Union, with the Nordic countries (Denmark, Finland and Sweden) and Austria utilising in excess of 5 percent of arable land. Greece, Portugal and Spain show high growth rates from a low base. The fastest contribution to total European Union growth in recent years has come from Italy and the United Kingdom, as these countries have a much greater land base than most of the northern European Union countries. Changes in government payments to organic farmers raised organic production significantly in these countries.

There is also a new category of organic farmers: subsidy farmers. In countries such as Austria, Finland, Italy and Sweden a high proportion of the organic land is managed by farmers that do not sell their products as organic. For example, in Sweden, the organically managed land represents 14 percent of total agricultural lands. However, only 7 percent of this area is certified organic. Sweden is the only country in the European Union where organic farmers can receive subsidies without being certified as organic. In Japan, according to an undated International Trade Center report, the area of organic production in 2000 amounted to a mere 1000 ha, just 0.02 percent of the total area under production. Another estimate suggests 5000 hectares. These low estimates are consistent with estimates of imports accounting for 80-90 percent of market share.

ITC 20 reports that approximately one hundred developing countries produce certified organic products in commercial quantities. Land area under organic management in Africa, Asia and Latin America is currently small but increasing rapidly. A prime example is Argentina, where organic production grew from 231000 ha in 1997 to 3200 000 ha in 2001, representing a growth of the average sale of organic products of 90 percent per year. Likewise, land under certified organic management in Africa is expanding. The largest areas are in Uganda (122000 ha) and South Africa (45000 ha).

An Expanding Retail Market

In the absence of data on production, perhaps the best data to judge the size and growth of the organic market are retail sales. ITC has estimated retail sales of organic foods and beverages in the largest markets at almost $16 billion in 2000, with $7 billion in Europe and $8 billion in the United States. The Japanese market is dominated by imports, with perhaps 80 to 90 percent of the market supplied from overseas. The US Department of Agriculture reports that organic sales in Japan reached $3.2 billion in 2000 but perhaps only ten percent of the market, $300 million, was truly organic. The market appears to be growing strongly (15 percent) from a low base, although some exporters report confusion in the market since the introduction of the Japan Organic Standards in 2001 and a 90 percent fall in organic exports to Japan.

It appears some retailers found continuity of supply a problem with the higher standards and dispensed with organic sales altogether, choosing instead to focus on "chemical-free" products. Many of these had previously been labelled as "organic" and, after the introduction of the standards, were re-labelled and promoted as "green". As a further barrier, organic fruit imported into Japan risks being randomly fumigated. If this occurs, it can no longer be sold as organic. Such impediments limit development of the organic market in Japan.

ITC has estimated retail sales of organic foods and beverages in the largest markets at almost $16 billion in 2000, with $7 billion in Europe and $8 billion in the United States. The United States of America, the world's largest single market worth US$8 billion in 2000, expects a strong growth in organic retail sales, perhaps reaching $20 billion in 2005. The current estimate represents growth of 23 percent per year over the previous ten years. The growth in recent years is attributed to concerns about genetically modified organisms and supermarkets adding organic lines to their shelves.

There is some evidence that organic premiums in the United States are as high as in Europe. Premiums of 100 percent for several major fruits and vegetables and milk products were consistently observed throughout the 1990s. Kortbetch-Olesen, 2002, however, suggests consumer premiums are generally within the range of 10-20 percent, although exporters may occasionally receive 100 percent premiums if supplies are short. In the United States markets, consumers appear more concerned about the environment than is the case in Europe, where food safety and health seems more of an issue.

The US Department of Agriculture predicts organic agriculture will remain an expanding sector for some time. In contrast to Europe, in the past the US taxpayer has not subsidised organic growers specifically, although some states provide financial assistance with certification. This is likely to change with the US 2002 Farm Bill, which provides for nationwide organic certification cost-share assistance. In the European Union, food retailing chains or processing companies are active in increasing the supply of organic products in order to improve their competitiveness by responding to growing consumer demand for organic food.

Wholesalers or processors offer producers long-term contracts which guarantee a price premium for organic products over several years. Price premiums for organic commodities are 20 percent lower in countries with a high involvement of general food retailers, mainly due to lower distribution costs.

In 2000, The European Union was a net importer of organic cereals, oilseeds, vegetables and meat and a net exporter of organic olives, wine and dairy products. Price premiums varied, in 2000, from 31 percent for organic red table wine up to 113 percent for organic chicken. Countries which have the highest market share by volume for organic products have a common national label and consumer recognition of this label. The role of big food industries in boosting organic agriculture is further illustrated by the New Zealand dairy sector. The entry of New Zealand Milk Products, Fonterra's manufacturing and ingredients marketing division, in September 2002, into processing organic milk to produce organic cheddar cheeses will change the composition and quantity of organic exports. The decision to enter the organic market has come after extensive market evaluation to ensure that market premiums of at least 20 percent are sustainable for the long-term, in order to match the higher costs structure.

In most developing countries, there are no markets for certified organic products. In some countries, however, organic urban markets are developing. Expanding demand for organic foods in developed countries is expected to benefit developing country exports, via new market opportunities and price premiums, especially for tropical and out-of-season products. Developing country exporters, however, will need to meet the production and certification of those in developed countries and overcome consumer preferences for local production.

SALES OF ORGANIC AGRICULTURAL PRODUCTS

Variations in Market Share

In spite of high growth rates, sales of organic agricultural products in industrialised countries in 2000 are estimated at less than 2 percent of total retail food sales. However, in particular countries and for particular products the market share of organic agricultural products can be appreciably larger. Organic horticulture products in British Columbia, Canada, occupy 8.7 percent of commercial sales. Organic food sales in Denmark occupied 3 percent of the market. In Germany individual commodities such as organic milk products had over 10 percent market share and organic baby foods 80-90 percent. Organic coffee accounts for 0.2 percent of world coffee consumption and up to 0.5 percent of the United States and European Union coffee markets.

Organic sales have stalled in some markets or for some products in countries which have previously experienced sustained growth for several years. This indicates that markets do not expand continuously and may reverse due to changes in supply raising or lowering prices, policies and promotion effects, or plain fickleness on the part of buyers. Clearly, different markets are being driven by different factors. Growth of organic vegetable consumption flattened out in the Netherlands in 1997, yet doubled in Belgium, Greece and Switzerland. Cereal sales also fell in the Netherlands from 1993 to 199736 while growth continued strongly in Finland.

In the United States, the organic sector grew rapidly between 1972 and 1975, collapsed, and began to slowly grow again between 1978 and 1981. Between 1981 and 1983, the United States organic sector was not exempt from the chain of bankruptcies that affected the whole agriculture sector. Combined to USDA's unfavourable policy towards organic agriculture, the infrastructure for producing and marketing organic commodities took almost the entire decade to be rebuilt to the 1980 levels. There seems no obvious relationship between share of organic production as measured in area under organic management and the share of the organic market, in Europe at least. Market shares for milk products vary tenfold between Sweden (2-

3 percent) and Finland (0.2-0.3 percent), yet the share of land under organic management is similar.

Organic Market Transparency

ITC reports that the "organic market transparency is far from optimal": there is need for data compiled over several years in order to identify growing markets and forecast market development. As markets mature, it is reasonable to expect information to flow quickly through the markets, implying that producers, distributors and consumers respond rapidly to changes. This is evidenced by the absence of large price shocks and of regional price differences between markets. For storable or easily transportable commodities, such as wheat and rice, one would expect relatively similar prices across the single market of Europe. In the organic sector, this seems to hold true for milk products, with premiums across Europe of around 20-30 percent, but not so for the other major products.

As the organic market matures, it should become more efficient. Oversupply of organic products cannot be absorbed by markets unless all market actors work together to benefit from the economies of scale. Larger volumes of organic products should lead to a lowering of the organic premium, in particular by reducing transport costs and the marketing margin. As the organic market matures, it should become more efficient. Oversupply of organic products cannot be absorbed by markets unless all market actors work together to benefit from the economies of scale.

High Marketing Margins

Marketing margins represent a stumbling block to future rapid expansion of the organic food sector. Traditionally, organic products were sold through alternative channels, that is, farm-to-home deliveries, farm gates, open-air markets, specialty and health food stores rather than supermarkets. Some consumers enjoyed the feeling of obtaining their food directly from the producers, almost as if they had grown it themselves, and producers had opportunities to capture some of the marketing margins.

With increased distance between producers and consumers and the entry of traders and retailers in the organic sector, organic marketing margins raised. There are of course reasons for the high margins. Margins are payments for storage, transport, handling, packaging, processing and insurance. With organic producers comprising a small proportion of the agricultural industry, individual producers are dispersed, adding to the costs of transport. Storage may be more expensive without pesticides. The need for segregation may increase costs of handling and require individual packaging and labelling. Sometimes the absence of infrastructure meant producers had to perform some of these handling functions themselves.

Small Economy of Scale

There appears to be significant scope for a reduction in marketing costs as the organic market grows and matures. Economies of scale should reduce these costs to near the margins for conventional products. If marketing costs could be reduced to conventional levels, additional costs of production of 10-20 percent, perhaps even 50 percent, would almost be unnoticed by consumers. This could result in a rapid expansion of organic market share.

Table 1. Producer and retail premiums in Denmark, 1997-1998

Organic product	Producer premium (%)	Retail premium (%)
Vegetables	25-50	20-50
Cereals	60-70	0-20
Dairy products	20-25	20-30
Potatoes	25-50	20-50
Fruit	>100	50-100

For example, conventional produce for which the farmer receives US$100 is sold to the final consumer for US$200. Organic premiums of 25 and 20 percent for the producer and consumer respectively imply that the organic producers are receiving US$125 (=US$100*125 percent) and the organic consumers paying US$240 (=US$200*120 percent). Hence, the organic marketing margin is US$115 (that is, US$240-125) compared with the conventional margin of US$100 (that is, US$200-100). Where the marketing margins are high, for products with a high degree of processing for example, the retail premium should be relatively low. This is the case with Danish cereals (Table 1). Where the retail premium is zero, the extra production costs are absorbed before getting to the consumer.

Minimally Processed Foods

Consumers of organic produce have traditionally expressed a preference for minimally processed foods, and organic standards have reflected this to some extent, encouraging minimal levels of processing. To date, the share of processed products and the degree of processing is rather low, with perhaps dairy products, beverages and cereals the exceptions. This low level of processing can be attributed to traditional values as to what organic is, and concerns about the need for, or the absence of, additives to preserve the product.

Limited organic food processing constrains the range of products that can be sold. As the organic industry develops and moves to mainstream markets, there is pressure from some producers and consumers to increase the level of processing. The development of supermarkets, often with vertically integrated systems from producer to consumer, has increased the scope for processed organic foods. These entities are more likely to seek raw materials from abroad if this is found to be advantageous. There is an increasing trade in bulk products from developing countries to developed countries for further processing.

Higher degrees of processing means that for many organic products, such as fruit and vegetables, the superficial blemishes are less important. This trend towards industrialisation of organic agriculture is of course resisted by many in the organic community who see it as a move away from traditional values. The size and nature of the certified organic market in the future depends somewhat on how this issue is resolved.

INTERNATIONAL TRADE CHARACTERISTICS

Trade Flows

Although there are no official data on trade in organic products, evidence suggests that, with demand outstripping production, trade in organic products in developed countries is relatively significant. Much of the organic trade occurs intra-European Union and intra-NAFTA countries. Much of the inter-continental organic trade is between the United States and Europe. About one third of the United States organic production is exported, mainly to Canada, Europe and Japan. As already seen, 80-90 percent of organic produce consumed in Japan is imported. About 70 percent of the United Kingdom consumption is imported, mainly from other European Union members.

Increasingly, developing countries are managing to meet the necessary standards and supply tropical and out-of-season products. Export markets are the major outlet for many developing countries' organic producers, given the small size of local markets. Egypt and Morocco have the advantage of proximity to the rapidly expanding European market that seeks fresh products such as oranges all year round. Argentina has third-country status with the European Union, reducing the costs of re-certification and inspection.

China has the potential to be a key developing country exporter, with a large agricultural area, low labour costs and proximity to the Japanese market in particular. In future, countries joining the European Union will obviously have increased opportunities to export to current members. Chile is increasing its export of counter-seasonal horticulture products to North America. Increasingly, developing countries are managing to meet the necessary standards and supply tropical and out-of-season products.

MARKET ACCESS

In terms of market access, organic products do not per se receive preferential access over conventional products, and there are significant tariffs (of 100 percent or more) on many agricultural products in most developed countries. However, many agricultural tariffs have a specific component (that is US$/t) so high-value goods face a lower percentage tariff. For example, imports of rice into Japan attract a duty of 402 yen (US$3.07) per kg. In percentage terms, this represents a higher duty on conventional rice compared with the higher-priced organic rice. As a result, such specific tariffs favour the import of organic products.

Since 1995, many agricultural products are imported under tariff rate quotas, with in-quota imports attracting a nominal tariff of perhaps 5 percent while imports outside the quota may face tariffs of 50 or 100 percent or more. At these levels, the tariffs may actually prevent any over-quota imports from coming in, which is often the intention. Quota restrictions such as these should favour high-value products, such as organic goods. This is because the quotas are based on physical quantities, and a greater value of exports can be supplied if the per unit value is higher.

However, what actually happens depends on how the quotas are administered. Many are allocated on a historical basis to countries, which have exported in the past. This makes it difficult for countries without a history of exports to break into the market. The second common method is first-come-first-served. This means that quota is allocated to those who ask in sequence. This favours countries with the ability to export early in the accounting period. It is not clear whether this favours organic exporters or not. Therefore, changes in the way in which imports are regulated may affect different crops and different countries in different ways, some working in favour of organic producers, some against.

ORGANIC AGRICULTURE GROWTH

Trends in Market Growth

Most analysts looking at the long-term future of the organic industry are cautiously optimistic, predicting annual growth of organic markets around 20 percent in the medium term. The market for certified organic food in 2000 was around US$16 billion, assuming the Japanese market is a conservative US$300 million.

The lower estimate of market growth rate for individual countries is different for each country and between 10 and 25 percent. Assuming the lowest growth estimates (10 percent for each country) are maintained for ten years, it would boost the global market to US$61 billion in nominal terms in 2010. This would represent a market share of 3.5 percent, assuming just 2 percent growth per annum in the total food market. Likewise, the higher estimated growth rates (between 15 and 30 percent), if sustained in all organic markets, would see retail sales reach US$94 billion by 2010 with a market share of 5 percent (Table 2).

Table 2. Estimates for organic food and beverages retail sales in 2010

Country	Expected medium in 2010* (US$ m)	Estimated sales term growth rates (%)
Germany	10-15	5 706-8 900
U.K.	25-30	9 313-13 786

Italy	15-20	4 046-6 192
France	15-20	3 034-4 644
Switzerland	15-20	1 719-2 631
Denmark	10-15	908-1 416
Austria	10-15	648-1 011
Netherlands	10-20	584-1 393
Sweden	20-25	774-1 164
Other Europe	10-15	778-1 214
Sub-Total (Europe)	-	27 510-42 351
U.S.A.	15-20	32 364-49 534
Japan	10-15	778-1 214
Australia	10-15	441-688
New Zealand	10-15	153-239
Argentina	10-15	52-81
China	10-15	31-49
Taiwan	10-15	26-40
Philippines	10-15	16-24
Total	15-20	61 372 - 94 220

Some authors have speculated about growth rates as high as 30 percent per annum, and indeed these have been observed in some markets. A sustained 30 percent across-the-board growth would see organic sales scale US$210 billion in 2010, or 11 percent of the global market. This seems unlikely given the current state of the market. If the high growth rates apply until 2005 and the lower estimates thereafter, the global sales in 2010 would amount to US$73 billion. The lower rate, from 2006, has a greater influence on the final estimate because the base is higher. These ten-year projections serve to illustrate the type of growth rates needed for the markets to attain certain levels or shares.

It is clear from the variation in the 2010 projections that the assumed growth rates make a lot of difference to sales ten years ahead. If organic sales have been unusually rapid in recent years because of food safety concerns from conventional foods, it could be that lower growth rates are more likely in the latter part of the decade. Scandals and food scares could also reduce demand for organic food in cases of:

— widespread and repeated fraud leading to consumer cynicism and scepticism of the organic label;
— a food safety threat does not give organic foods an obvious advantage over conventional foods.

Product Demand

Factors that affect demand for organic products in the long run are income growth, relative prices, environmental consciousness and exogenous shocks. In some developing countries, local markets for certified organic agriculture are emerging, especially through specialised shops in large cities. Although these markets are still very small and not assessed, they could develop quickly. Local organic markets tend to increase with income growth and heightened environmental consciousness.

Theoretical and empirical evidence suggests that demand for environmental goods is positively related to income levels once national incomes rise beyond US$8 000 per person per year. This varies across countries and types of problems, with local pollutants of air and water addressed first. An exception seems to be global pollutants, such as greenhouse gases, but pollutants involved in conventional agriculture are among those that tend to be reduced as the higher-income countries impose more stringent regulations. To the extent that the demand for organic food is based on improving the physical and social environment, rising income levels should increase the market share of organic products.

The most rapid increases in income growth are occurring in areas where the organic market share is low or the product is uncertified. China is a large and rapidly expanding market, but organic sales are starting there from a low base and are often overridden by "green" foods. Not all the evidence confirms the relationship between income and organic food demand. An attitudinal survey of 1 200 Australian consumers found that the number of people consuming organic food tended to increase with income, but not very significantly.

In Argentina, about 20 percent of the total value of national organic production was sold on local markets in 2000. The 2001, the economic crisis lowered the internal organic market to 7 percent. It is however interesting to note that this reduction of internal organic markets affected high value animal products (poultry, beef, milk) but not the internal market of vegetables and eggs and honey, which proportionally, increased.

An attitudinal survey of 470 Kansas (USA) consumers found that income, relative prices and family size determined organic purchases. The authors concluded that if certain marketing barriers could be surmounted, significant increases in sales of organic products would be possible as incomes increase and, more importantly, relative prices move in favour of organic products. It is instructive to examine a mature, well-functioning organic market. Denmark is one such example. The organic agriculture labelling system is well known and trusted.

Most of the products are sold through supermarkets. Supply is relatively stable and the price premiums are moderate. A study of 2 000 Danish households shows the estimated responsiveness of consumers to changes in retail prices.

The estimates suggest consumers are very responsive to changes in organic prices, especially for livestock products. The analysis suggests that if organic prices fell by 20 percent, the organic market share would rise from 10 to 15 percent for dairy products (Table 3). This implies that a 20 percent price reduction elicits a volume response of 50 percent.

Table 3. Estimated changes in Danish organic market shares from 20 percent reduction in prices

	Existing share (1997-1998)%	Estimated market share %	Volume response %
Dairy products	10	15	50
Bread and cereals	5	7	40
Meat	1	2	100
Other	4	6	50

While few markets share the characteristics of Denmark, the implications of these results for market shares are obvious. Lower relative costs to the final consumer will boost the market. Changes in tastes and preferences are difficult to analyse or predict, but are probably the driving forces behind the present growth in demand for organic commodities. Increasing awareness of environmental and food safety issues are a major factor. Concern over animal welfare is a major force in some markets.

In recent years concerns about genetically modified organisms have helped increase awareness in some countries, as have concerns about livestock diseases and antibiotic-resistant bacteria. If new technology proves to have health or environmental risks, the demand for organic products is bound to increase. On the other hand, if the current concerns abate, the recent growth in demand for organic products may slow down.

Supply of Organic Products

Several factors constrain the ability to increase supply of organic products in response to growing demand, at least in the short term. On the other hand, there are a series of factors that should lead to a lowering of costs of production, and may well be much more important than the negative ones. It seems reasonable to expect that those who can most readily convert to organic agriculture are likely to do so first. Complete newcomers will take longer to come on stream. Organic management requires not only a different mix of inputs and outputs but also a new way of doing things. Knowledge on how to farm organically is not readily available, as in most countries few courses or advisors are available to teach those who are interested. To some extent, farmers have to learn on the job. Those who do not have this aptitude will find the transition more difficult and expensive.

Factors that affect demand for organic products in the long run are income growth, relative prices, environmental consciousness and xogenous shocks (e.g. food scares). The increasing number of organic farmers and e-communication facilitates knowledge sharing and learning from one another's mistakes. As knowledge concerning organic practices becomes widespread, production efficiencies should improve. Increased acceptance of organic management methods may also lead to an increase in investments in organic agriculture research, escalating the productivity.

The conversion period is costly because of inevitable management errors, but also by additional investments, for example:

— new fencing may be needed to arrange smaller fields so that livestock can be used more strategically for weed control;
— silos and/or storage bins may need to be built on the farm;
— housing for animals may need to be redesigned, and built, to accommodate different needs, such as free-range arrangements;
— for some products (such as fruit), organic pest control is so difficult that it might entail replanting the whole orchard to more resistant varieties; and
— organic farmers may need to establish new marketing

strategies instead of through the established marketing channels.

All these factors raise costs of production. Of course, these difficulties are not the same for all enterprises. It is relatively easy to convert dairy farming to organic production in Austria, Denmark, Finland and Switzerland, and this has led to a significant penetration of organic produce in the milk and dairy product market in these countries. In addition to the knowledge and investment required during the conversion period, organic farmers have temporal requirements, especially in terms of the time it takes for soil to respond to new management, with possible consequences for lower yields in those years. Also, products grown during the conversion period often do not qualify for price premiums.

Another constraint is the rising cost of inputs as demand increases. Although organic farmers prefer to rely on a closed system with minimal use of purchased inputs, this is not always possible. One concern is the availability of seed or plant material acceptable under organic management (such as GMO-free seeds, seed not coated with a fungicide or suitable open-pollinated varieties). Another is the availability of suitable livestock breeds. Insufficient quantities of organic nutrients could cause problems in organic cropping systems which are not integrated with livestock or which have a relatively intensified production. Sources of nutrients include animal manure, algae extracts, rock phosphate, nitrogen-fixing crops and composted household and food processing waste. New technologies may need to be developed to make maximum use of these, and other potential inputs. Some input costs may fall as demand for the input increases and boosts production.

With more specific reference to livestock, production standards allow zero or only a small proportion (less than 20 percent) of feed and fodder to be non-organic. The gradual decrease in percentage of non-organic feed will increase the price of organic feed if supply does not respond timely to the increased demand. According to the Danish Ministry of Food, Agriculture and Fisheries, supplies of organic broiler chicken were limited by insufficient availability of organic feed.

Pig production faced similar constraints. However, other factors also play a role in the determination of the price. Increased feed prices will mean that some farmers will reduce livestock production, and some will increase feed production. Modelling of the Danish organic sector suggests that lowering the allowable proportion of non-organic feed on all the different types of farms from 20 percent to zero changes national net returns from DKr 3 566 million (approximately US$471 million) to DKr 3 350 million (approximately US$443 million) for the sector as a whole assuming 20 percent of all farms are using organic management techniques.

The pig sector is hardest hit, with national net returns DKr 2 447 million (approximately US$323 million) to DKr 2 293 million (approximately US$303 million). The relative effects would obviously be much greater with wider levels of adoption of organic methods—at least under present conditions of input and output prices, and the state of technology. These results cannot be generalised to other countries with a different industry structure than Denmark, or indeed to different years.

Organic agriculture represents a different paradigm from conventional agriculture. Techniques and innovations applicable to one may not be readily transferable to the other. Increased research into organic methods is likely to significantly reduce the cost disadvantage that acts as a barrier to many farmers converting to organic agriculture. The increased demand for organic nutrients and fodder will tend to increase costs in the short term as organic agriculture expands, depending on the particular country and industry.

The availability of organic seeds and livestock breeds may also constrain growth in the short term. However, at the levels of adoption considered likely in the next ten years, supply constraints of this nature are unlikely to be binding in most markets. The various factors pushing costs down are likely to be of greater significance. The costs of supplying organic products to the consumer are likely to fall as economies of scale and research and development in production and marketing bring about greater efficiencies. The costs of supplying organic products to the consumer are likely to fall as economies of scale and research and development in production and marketing bring about greater efficiencies.

ORGANIC AGRICULTURE STANDARDS

In addition to underlying demand and supply factors, a further issue determining the future of the organic agriculture market relates to production, processing and handling standards as well as to compliance with the organic guarantee system (inspection, certification an accreditation). There is a trade-off between breadth and depth in the setting of standards. Clearly, as with any product, the more stringent the standards are, the more closely some consumers of the product will be provided with the exact characteristics they demand. More stringent production standards, however, limit the size of the organic supply, as fewer farmers will be in position to participate.

The European Commission is under permanent pressure from the food industry to make the organic standards more flexible: the risk is that the concept of organic agriculture may become diluted. The organic sector is in competition with other sustainable agriculture alternatives, all clamouring for the attention of the consumer, and may lose out to the alternatives if consumer expectations are unmet.

Standards involve compliance, ensuring that the organic standards are met. International organic markets (European Union, Japan, the United States) are now all subject to government rules and conditions of access. Developed countries face problems of multiple accreditation of certification bodies and administrative complications. Developing countries often require certification by a foreign certification body, which can be prohibitively expensive. Some countries manage to have their certification body accepted as equivalent to that of the importing country.

The harmonisation and equivalency of organic production and certification standards can facilitate trade as they lend credibility to the product. It saves on costs and removes the disincentives currently preventing the participation of many producers in international organic trade, especially smallholders in developing countries. The evolution of organic agriculture standards and the establishment of guarantee systems that cater for diversity of conditions and needs today are likely to influence the nature and size of the organic market and its position relative to competitors in the years ahead.

Governments in Organic Agriculture

European union

The role of most governments in organic agriculture has been regulating organic production and certification and supporting producers. In 1997, funds to support organic agriculture amounted to Euro 186 million, less than 0.5 percent of the total Euro 40.5 billion (US$40.5 billion) annual agricultural budget of the European Union. The evolution of organic agriculture standards and the establishment of guarantee systems that cater for diversity of conditions and needs today are likely to influence the nature and size of the organic market and its position relative to competitors in the years ahead.

The European Union Agenda 2000 policies increased support for the provision of environmental services without specifically assisting organic agriculture. Perhaps most important of all, changes introduced in Agenda 2000 allow member states to switch direct support from large farms to agri-environmental measures, putting the onus on national governments to determine their environmental outcomes and their support for organic agriculture. However, organic agriculture has to compete with other agri-environmental schemes, as farmers can receive government funding by choosing to comply with alternative schemes such as integrated pest management. Nonetheless, there is evidence of greater funding and greater uptake of support to organic agriculture. For example, assistance for farmers converting to organic agriculture in the United Kingdom increased from £1 million (approximately US$1.5 million) in 1997 to £18 million in 2001.

A move towards organic agriculture would reduce European Union support payments but increase government outlays in the member states. The reduction in European Union payments comes about because livestock are less heavily subsidised than cereals and organic producers tend to have more livestock in their system. A shift towards organic methods would most likely lead to a fall in production of cereals and reduce the need for intervention and storage costs and exports subsidies. A detailed study of the Danish agricultural sector, using data collected by the Danish Institute for Agricultural and Fisheries Economics, looks at the contribution of domestic support to gross returns under the arrangements that existed at the time. In 1996, farm subsidies averaging DKr 149 000 per farm (US$17 890 at current exchange rates) exceeded net returns and contributed 9 percent to gross returns on conventional farms.

The European Union strategy on sustainable development tacitly recognises the impact of the Common Agricultural Policy (CAP) on the environment (e.g. decline in biodiversity). In May 2001 and March 2002, European ministers of agriculture met to discuss organic agriculture prospects. A commitment was made to launch a process towards the development of a European Action Plan within the next two years containing a consensual strategy covering all aspects of organic food and farming in Europe. The Action Plan will ensure, among others, that organic agriculture is a key element of the Common Agricultural Policy reform and environmental programmes and that it will take into account other policy initiatives. The establishment of such an Action Plan on organic agriculture in Europe is likely to have a multiplier effect on other countries.

Several European governments have set objectives for conversion of agricultural lands to organic management, with a view to achieving environmental or other objectives. By 2005, the target for conversion to organic land for Denmark and Italy is 10 percent each and of 20 percent for Sweden. By 2010, the Netherlands and Norway target 10 percent each, Germany 20 percent, and the United Kingdom 30 percent.

Countries in the European Union with the largest area of agricultural land (France and Spain) have very low rates of conversion at present but Spain has recently boosted its organic conversion. Attaining these ambitious targets depends on government funding and consumer support. Targets and subsidies are not uniform across the European Union member states, reflecting differing objectives. Policies for conventional agriculture also affect organic agriculture. Chiefly, building a domestic consumer base is of fundamental importance for any country to foster organic conversion.

United states of America

In the United States, there has been virtually no federal-funded support for organic agriculture. The Conservation Security Programme provides individual farmers who can meet specific environmental requirements with up to US$50 000 per year. Although organic farmers are well placed to meet these requirements, few of them are large enough to qualify for this amount. Using reasonable assumptions about average payments (between US$5 000 and US$15 000 per farm per year) and growth rates in the number of farms (of between 14 and 17 percent per year), discounted costs would most likely be between US$343 million and US$1 031 million. As less than 50 percent of United States organic farmers earn US$15 000 or more per year, these proposed payments represent a substantial subsidy.

The 2002 US Farm Bill includes funds for organic certification cost sharing which cover 75 percent of an operation's inspection and certification fees, up to US$500 per farm per year. In some states, such as Minnesota, converting lands have started receiving organic transition support payments of US$20 per ha for crops and US$10 per ha for pastures.

Agricultural and Environmental Policies

Agricultural and environment policies are linked. Organic agriculture is viewed by some governments to have a specific role, with fewer negative environmental impacts and greater positive ecological services than conventional agriculture, notwithstanding that there is a range of conventional farming systems with varying degrees of environmental impact.

Government funding for organic production in most countries has evolved from agri-environmental schemes. Environmental objectives will continue to drive government support towards organic agriculture. For example, the Organic Federation of Australia has developed an ambitious five-year plan that aim to raise the organic sector from 1 to 4 percent of agricultural production by 2006. The Government, however, would consider the sector within the commodity-based environmental management systems currently under development.

Organic agriculture would receive a major boost if the environmental externalities associated with conventional agriculture were charged to those contributing to the problem. This has proved difficult in agriculture because the source of pollution is often dispersed and dependent upon a variety of factors and difficult to quantity. Nonetheless, moving to producer-pays principle would mean that the cost of cleaning up the pollution from agriculture would shift from taxpayers to agricultural producers. These costs are then likely to be passed on to consumers, with a favourable impact on the relative price of organic and conventional goods.

The contribution of organic agriculture to the implementation of international environmental agreements has not yet been sufficiently assessed. There are, however, reasons to believe that organic agriculture could facilitate reaching environmental targets, such as:

- the Convention on Biological Diversity targets for conservation and sustainable use of biodiversity;
- the Kyoto Protocol on Climate Change targets for greenhouse gas emissions and eventual agreement on the role of sinks to sequester carbon;
- the Convention on Combating Desertification objectives for rural development.
- Organic agriculture might also contribute to other policy objectives by providing food perceived to be safe

and healthy by some consumers, reducing government payments relating to surplus production, furthering regional development and raising animal welfare.

Prospects for Organic Agriculture

In industrialised countries, demand for certified organic products is expected to continue to increase, perhaps above the 20 percent growth rate registered in current years. However, the future growth of certified organic agriculture will depend more on supply constraints than on changes in demand. The tendency so far has been for demand to grow faster than supply, but this may moderate as more "followers" (as opposed to "innovators") enter the market.

Developing countries are just starting to benefit from organic market opportunities but under present conditions, large producers and operators are better placed to access international markets. The limited quantities of organic products and the demanding quality standards and rules governing organic production and processing might limit the extent to which developing countries can fill the demand for organic food in northern markets. Organic food trade might be discouraged by difficulties in complying with foreign standards as well as costly certification systems, especially if international equivalence is not established. Access to inspection and certification, and the need to develop new ways in processing organic food, are major challenges that large and established food companies are better placed to handle.

Future growth of certified organic agriculture will depend more on supply constraints than on changes in demand. Multinational food companies are expected to be major players in organic food supply, both in terms of contracting production and international trading. In particular, the growth of processed organic foods will be facilitated by these companies' capacities to assemble ingredients from different parts of the world and to guide production to meet their specific needs. At the same time there are numerous opportunities for developing country producers and exporters to enter markets for value-added organic products using simple available technology.

As world agriculture globalises, few and large private companies will increasingly control world food supply chains. A handful of multinationals not only has the potential to limit farmers' choice on the type of food they produce but can also influence consumers' preferences through massive opinion-forming campaigns. Two contrasting forces will influence the growth of organic agriculture. On the production side, multinational agrochemical companies, now also providers of patented engineered seeds and breeds, are unlikely to accept the loss of a large share of the inputs market; large investments are already made in campaigns to deter producers and consumers of organic foods.

On the demand side, mega food distribution companies will attempt to guide and mould the organic market to their mass marketing requirements as they have done in the conventional sector. Large retailers have started establishing their own organic standards, based on minimal input substitution requirements, in order to fit the industrial production and distribution model.

However, different factors influence different types of consumers. Large food distributors are realising that nor all consumers can be expected to spouse the "fast food" culture, neither can all consumers be expected to enter the organic food culture. Within the organic consumer category, there are those who demand organic convenience food. Most of the demand for organic food, however, is based on minimally processed foods, the rejection of mass-market trends and on increasing fragmentation of markets.

The growth of industrial corporate organic foods is increasingly triggering responses of the organic agriculture community, mainly to maintain the integrity of organic agriculture and the viability of small organic farmers. Local markets, farmers markets, home delivery and community-supported schemes are emerging as alternative marketing strategies for organic agriculture.

The rediscovery and valorisation of regional food production and local culinary cultures, coupled with energy concerns, might limit future organic food trade essentially to unprocessed tropical products. In fact, high-energy consumption entailed in international food transport and the likely introduction of "food miles" standards (related to the distance between production areas and markets) for certified organic food might result in the establishment of narrow food chains that will strengthen domestic organic markets.

The most-traded organic products today are cereals, fruit and wine, as well as tropical and out-of-season fruits and vegetables. In the coming years, developing countries are likely to gain a relatively substantial market share for organic coffee, tea, cocoa, herbs and spices, dried fruit and nuts. Organic baby food, beverages and frozen vegetables will likely dominate imports.

Oilseed trade (especially temperate oilseeds such as soybeans and rape) is subject to major changes as oilseeds are the focus of biotechnology. Future evidence on the food and environmental safety, and of consumer acceptance, of genetically modified oilcrops might either increase the potential of net exporting countries through improved conventional production or create new markets (and exporters) of organic oilcrops. At present, major organic soybean producers (Argentina, Brazil, and Paraguay) are being heavily confronted with GM soy (and Bt corn) that have become mainstream in these countries. In Canada, organic producers can no longer market their rape as "organic" because of GM-rape contaminated organic fields.

Currently, trade in organic meat and livestock products (e.g. dairy) is low, partly because European Union-wide standards were adopted only in 2000 and because USDA lifted its prohibition of labelling and marketing organic meat only in February 1999. Given the high-value of livestock products as well as concerns with industrial meat safety, there is significant scope for organic beef and sheep meat exports.

World meat demand is increasing and is forecast to continue to increase with economic growth; many countries (such as Japan, the largest importer of meat, followed by the EU) are expected to seek more organic meat as a result of food safety concerns. Australia and New Zealand, who are the largest exporters of meat, have a comparative advantage for massive conversion to organic meat production.

It is expensive for intensive livestock producers to convert to organic production, but converting extensive, pasture-based systems could become economically more attractive if price premiums could be captured for organic meat and livestock products. In fact, the spectacular expansion of organic land in Argentina from less than 0.5 million ha in 1999 to 3 million hectares in 2000 occurred mainly on grassland, in response to demand for organic meat, although this reflects reclassification and extension of certification rather than real switches between farming systems. Organic livestock could improve both the productivity and profitability of existing pastures and put into sustainable use unexploited natural grasslands (for example, in China).

Organic livestock could improve both the productivity and profitability of existing pastures and put into sustainable use unexploited natural grasslands. Current agriculture policy revisions, as in the European Union, in favour of organic agriculture might result in smaller cereal surpluses. Should Australia and North America (which together with Western Europe are the largest cereal exporters) follow this trend, a deficit of organic feed availability on world markets might occur. This might induce changes in animal feeding practices in developing countries, which depend on imported cereals for feed.

An eventual decrease in the availability of imported feed will undoubtedly bring a period of difficult adjustment to developing countries. A positive adaptation will be the establishment of narrower food production-consumption chains through efficient use of local resources and more developed domestic markets for organic commodities.

Favourable policy commitments and related support programmes lead to assume that Western Europe, on average, could have a tenth of its total agricultural lands under organic management by 2010. If these policies and the organic growth trend were sustained, up to a quarter of the European Union agricultural lands would be organically managed by 2030. Lack of data and rapid changes do not allow predictions of the extent of organic land management elsewhere.

Land use changes will not be linear but will respond to technological innovations and to unforeseen events. An example is the development of organic agriculture in Europe: it took 30 years for organic agriculture to occupy 1 percent of agricultural land and have a visible place in food markets. Then, the food safety crisis resulted in a spectacular (and unforeseeable) increase, with countries such as the United Kingdom aiming at 30 percent of land being farmed organically within 10 years. This indicates that growth trends of organic agriculture are subject to major deviations.

Keeping in mind factors that might radically change the future direction of agriculture as a whole, the demand for certified organic products in industrialised countries and the urge to meet organic supply requirements by developing countries is expected to induce major land-use changes worldwide. This would consequently bring changes to the food supply and distribution models in both developed and developing countries.

The demand for certified organic products in industrialised countries and the urge to meet organic supply requirements by developing countries is expected to induce major land-use changes worldwide. For example, societal demands on the quality of land and water and the expansion of organic land management will entail changes in spatial policies in order to avoid contaminant drifts from neighbouring areas farmed conventionally in order to protect organic farms or their collocation in suitable and protected regions. Adoption of organic livestock production could reverse the on-going trend of intensive animal production systems that create considerable pollution problems as well as off-site degradation where feed and fodder are produced.

More specifically, organic agriculture could be a major force towards:

- diversification and intensification of existing agricultural land use through improved yields and increased cropping intensity. In East Asia, for example, organic rice paddies have huge potential not only for improving rice productivity but also for double cropping by integrating fish;
- building soil fertility and preventing erosion and, in consequence, decreased annual loss of agricultural land to degradation and desertification;
- decrease in the clearing of new land for agriculture;
- achieving good use of land set aside;
- producing food in many protected areas and especially in buffer zones;
- providing an option for the productive use of available urban land for food production.

Considering the inevitable conversion of agricultural land to urban uses, the necessity to protect the environment and halt biodiversity extinction, and the importance of increasing food production without losing arable land or clearing new land, organic agriculture becomes an attractive option.

In most European countries, organic agriculture is perceived as a powerful means to recover and give new value and importance to the traditional characteristics of rural areas. Bio-ecological agritourism is being promoted as an activity that not only produces specific food and fibres but also offers typical landscape, environmental services and rural hospitality. There are also avenues for income generation and food production to be explored by developing countries that have a great environmental capital.

The certified organic food market comprises a small (US$16 billion) but growing share (1-2 percent) of the total food market. In Europe in particular, this growth has been enhanced, on the supply side, by production subsidies to organic farmers and, on the demand side, by adverse perceptions concerning the safety of conventional food spurred by crises of confidence. However, growth in the organic market has also been strong in markets where these factors have been less evident, such as Australia, Japan and the United States, and some developing countries.

It is likely that the global market for certified organic products would continue to grow at 10-20 percent per year for the foreseeable future. By 2010 the market may well have grown to US$61-94 billion in countries with certified organic markets, or between 3.5 and 5 percent of the global food market. This is in the currently certified markets; expansion will be greater if non-certified organic markets were included.

The deregulation of agricultural policies would leave decisions on the most suitable type of agriculture and food production to market forces. The growth of certified organic agriculture will be governed in the future according to market forces, specifically: the growing concern of society with the safety of food produced by conventional systems; the growing interest of multinational enterprises in organic products; decoupling of agricultural support from production; and the increasing emphasis on agriculture's role in providing public goods.

The implications for governments wishing to develop the organic sector include the development of the appropriate policy framework that ensures transparency and the appropriate quality of organic agriculture products. This is particularly important for exporting countries.

Over the medium-term, all forms of sustainable agriculture, including integrated production and organic agriculture, are expected to converge with regards to their contribution to environmental quality, food safety and fair trade. The distinction, however, will remain with biotech agriculture, namely genetically engineered food and fibres. The degree of success and acceptance of biotech agriculture will have a major influence on agricultural policies and the viability of organic agriculture.

Major forces influencing organic agriculture growth, including market dynamics, policies and technology development. By 2010 the market may well have grown to US$61-94 billion in countries with certified organic markets, or between 3.5 and 5 percent of the global food market.

REFERENCES

Altieri, M.A., "Applying agroecology to enhance productivity of peasant farming systems in Latin America", *Environment, Development and Sustainability*, 1: 197-217, 1999.

Avery, D., *Saving the planet with pesticides and plastic*, Indianapolis, The Hudson Institute, 1995.

9

Challenges of Sustainable Food Production

After half a century of high intensive input agriculture, the yield gap between best practices and farmers' fields remains large, agricultural lands continue to shrink and global environmental threats are a reality, e.g. erosion of biodiversity, desertification, climate change and other transboundary pollution.

Agriculture intensification contributes to over 20 percent of global anthropogenic greenhouse gas emissions. Agricultural activities affect 70 percent of all threatened bird species and 49 percent of all plant species. Uniform cultures have dramatically reduced the number of plants and animals used in agriculture; currently, 1 350 breeds face extinction, with two breeds being lost each week. Biodiversity erosion is exacerbated by the loss of forest cover, coastal wetlands and other wild relatives, important for the development of biodiversity and essential for food provision, particularly in times of crisis.

Despite FAO's and other institutions' efforts for global food security, the benefits of increased agricultural production often bypass the poorest sections of the world population. Over 450 million farmers have never had access to Green Revolution technologies such as mechanisation, irrigation, improved seeds and breeds, and synthetic fertilisers and pesticides. The World Food Summit stressed that hunger is a problem of access to means of food production or means to purchase food.

Although agriculture remains the world's single largest employer, rural economies suffer from sharp decline in real food prices and loss of entrepreneurial capacity. The terms of international exchange favour importations at the expense of local production. Developing countries, which traditionally have had a net surplus in agricultural trade, increasingly depend on food imports. The conventional food production model ties farmers into conditions of dependence on large corporations to buy agricultural inputs and to sell their produce.

Nature conservation: Protecting biodiversity at genes, species and ecosystem levels through germplasm banks and protected areas is not sufficient. The maintenance costs of gene banks are high, up to half of the material collected is in need of regeneration, and "freesing" genetic resources denies their evolution. Biodiversity is best maintained through sustainable utilisation and selection by food providers.

Animals move across boundaries and ecosystems are not immune to air and water pollution. The 12 percent of global land areas "fenced" for nature protection are located within or around a 40 percent land surface used by agriculture and forestry. Food systems should be viewed as an integral part of the ecosystem. "There is a need to manage agricultural land as part of a larger landscape that explicitly considers ecological functioning".

Rural livelihoods: Rural communities need land to derive livelihoods. National parks and protected areas have always had local culture as components of the landscape. Ecotourism management by local residents generates supplementary income for both management of protected areas and rural livelihoods. Agritourism creates new opportunities for farm employment, speciality foods and safe agro-ecosystems. There is a need to reconnect farmers to the land by linking their productive activities with nature conservation. A symbiotic relationship between agriculture and natural landscape exists in the presence of ecologically managed systems.

Food self-reliance: Poor farmers and market-marginalised communities need a food production model that relies on local natural resources and ecological management. Where food surpluses are generated, markets should guarantee "fair" prices. Managing local resources without having to rely on external inputs involves substituting purchased goods by knowledge of natural processes that optimise competition for nutrients and space between species within the agro-ecosystem.

By stressing diversification and adaptive management, agricultural systems can improve soil and water quality and the ecological services that support agriculture, while significantly decreasing vulnerability to weather vagaries or other factors. Where markets for ecologically-produced food exist, farmers' return on labour are higher through consumers' willingness to pay higher prices.

ORGANIC AGRICULTURE AND BIODIVERSITY

Organic agriculture offers a means to address food self-reliance, rural development and nature conservation. The common thread in this ambitious goal is the sustainable use of biodiversity; in terms of both agriculture contribution to biodiversity and biodiversity contribution to agriculture. To be successful, organic agriculture needs functional groups of species and essential ecosystem processes as its main "input" to compensate for the restriction on synthetic input use.

In fact, a close relationship exists between organic agriculture and the maintenance of biodiversity. This is expressed in the rules and regulations that govern certified organic agriculture and by the practical experiences of organic farmers around the world. Selected international codes and agreements expressing the organic agriculture biodiversity linkages and need for further consideration are the following:

- The Vignola Declaration and Action Plan on the Relationship Between Nature Conservation, Biodiversity and Organic Agriculture of the International Federation for Organic Agriculture Movements, the World Conservation Union and the World Wide Fund for Nature. This Declaration recognises that "organic agriculture is essential for conserving biodiversity and Nature" and "asks environmentalists, farmers, politicians, industry and international institutions to support and develop organic agriculture as the most ecologically-sound agricultural system".
- The FAO/WHO Codex Alimentarius guidelines for Organic Food define organic agriculture as "a holistic production management system which promotes and enhances agro-ecosystem health, including biodiversity, biological cycles, and soil biological activity. The primary goal of organic agriculture is to optimise the health and productivity of interdependent communities of soil life, plants, animals and people".
- The Convention on Biological Diversity (CBD) "encourages the development of technologies and farming practices that not only increase productivity, but also arrest degradation as well as reclaim, rehabilitate, restore and enhance biological diversity and monitor adverse effects on sustainable agricultural biodiversity" such as "inter alia, organic farming". Furthermore, the CBD Global Strategy for Plant Conservation includes outcome-oriented global targets for 2010 whereby using plant diversity sustainably considers, among others, organic food as an indicator of direct measures to reach the target of "30 percent of

plant-based products derived from sources that are sustainably managed".

— The Ninth Session of the Commission on Genetic Resources for Food and Agriculture recommended that FAO further studies the "contribution of genetic resources to the economies of member countries, the impact of agricultural subsidies on biodiversity, and the competitiveness of organic agriculture".

In organic agriculture, biodiversity is both instrument and aim. Natural ecological balance, below and above ground, is key to its success. A healthy soil is the base for food production and a diversity of plants and animals on land prevents pest and disease outbreaks. Although organic agriculture is committed to the conservation and enhancement of biodiversity, many systems today remain limited to input substitution.

To be unlocked, the real potential of organic agriculture on biodiversity requires a stronger shift to a systems approach, based on improved understanding of ecosystem functions. The presentation below describes the promising, but still scattered, results observed in organic agriculture systems. The food chain is described for soil systems, farming systems and the larger ecosystem. This involves descriptions of the impact of organic management on soil biodiversity, genetic resources for food and agriculture and wildlife biodiversity.

SOIL BIODIVERSITY

Soils contain enormous numbers of diverse living organisms assembled in complex and varied communities. Soil biodiversity reflects the variability among living organisms in the soil—ranging from the myriad of invisible microbes, bacteria and fungi to the more familiar macro-fauna such as earthworms and termites. Plant roots can also be considered as soil organisms in view of their symbiotic relationships and interactions with other soil components.

These diverse organisms interact with one another and with the various plants and animals in the ecosystem, forming a complex web of biological activity.

Environmental factors, such as temperature, moisture and acidity, as well as anthropogenic actions, in particular, agricultural and forestry management practices, affect to different extents soil biological communities and their functions. Soil organisms contribute a wide range of essential services to the sustainable functioning of all ecosystems. They act as the primary driving agents of: nutrient cycling, regulating the dynamics of soil organic matter, soil carbon sequestration and greenhouse gas emissions; modifying soil physical structure and water regimes; enhancing the amount and efficiency of nutrient acquisition by the vegetation; and enhancing plant health. These services are not only critical to the functioning of natural ecosystems but constitute an important resource for sustainable agricultural systems.

Capturing the benefits of soil biological activity for agricultural production requires adhering to the following ecological principles:

— *Supply of organic matter*: Each type of soil organism occupies a different niche in the web of life and favours a different substrate and nutrient source. Most soil organisms rely on organic matter for food; thus a rich supply and varied source of organic matter will generally support a wider variety of organisms.

— *Increase of plant varieties*: Crops should be mixed and their spatial-temporal distribution varied, to create a greater diversity of niches and resources that stimulate soil biodiversity. For example, diverse habitats support complex mixes of soil organisms, and through crop rotation or intercropping, it is possible to encourage the presence of a wider variety of organisms, improve nutrient cycling and natural processes of pest and disease control.

— *Protection of soil organisms' habitats*: The activity of soil biodiversity can be stimulated by improving soil living conditions, such as aeration, temperature, moisture, and nutrient quantity and quality. In this regard, reduced soil tillage and minimised compaction—and refraining from chemical use—are of particular note.

Improvement in agricultural sustainability requires, alongside effective water and crop management, the optimal use and management of soil fertility and soil physical properties. Both rely on soil biological processes and soil biodiversity. This calls for the widespread adoption of management practices that enhance soil biological activity and thereby build up long-term soil productivity and health.

Adaptation and further development of soil biodiversity management into sustainable land management practices requires solutions that pay adequate consideration to the synergies between the soil ecosystem and its productive capacity and agro-ecosystem health. One practical example of holistic agricultural management systems that promote and enhance agro-ecosystem health, including biodiversity, biological cycles and soil biological activity is organic agriculture.

Building of soil fertility is the cornerstone of organic agriculture. Organic practices create suitable conditions for soil biotic and abiotic resources through: manipulation of crop rotations and strip-cropping; green manuring and organic fertilisation; minimum tillage; and avoidance of pesticides and herbicides use. Scientific research in Europe has demonstrated that organically-managed soils significantly increase biological activity and total density and diversity of soil micro-organisms. Such biodiversity enhances nutrients recycling and soil structure. The impact of organic management on soil biological activity and related benefits is summarised below:

— *Abundant earthworms and arthropods*: Organic management increases the abundance and species richness of earthworms and beneficial arthropods living above ground, and thus improves the growth conditions of crops. The biomass of earthworms in organic systems is 30-40 percent higher than in conventional systems, their density even 50-80 percent higher. Compared to the mineral fertiliser system, this difference is even more pronounced. More abundant predators help to control harmful organisms. In organic systems, the density and abundance of arthropods, as compared to conventional systems, has up to 100 percent more carabids, 60-70 percent more staphylinids and 70-120 percent more spiders. This difference is explained by prey deficiency due to pesticide influence as well as by a richer weed flora in the standing crop that is less dense than in conventional plots. In the presence of field margins and hedges, beneficial arthropods are further enhanced, as these habitats are essential for over-wintering and hibernation.

— *High occurrence of symbionts*: Organic crops profit from root symbioses and are better able to exploit the soil. On average, mycorrhisal colonisation of roots is highest in crops of unfertilised systems, followed by organic systems. Conventional crops have colonisation levels that are 30 percent lower. The most intense mycorrhisal root colonisation is found in grass-clover, followed by the vetch rye intercrop. Roots of winter wheat are scarcely colonised. Even when all soils are inoculated with active micorrhisae, colonisation is enhanced in organic soil. This indicates that, even at an inoculum in surplus, soil nutrients at elevated levels and plant protection suppress symbiosis. This underlines the importance of appropriate living conditions for specific organisms.

— *High occurrence of micro-organisms*: Earthworms work hand-in-hand with fungi, bacteria, and numerous other micro-organisms in soil. In organically managed soils, the activity of these organisms is higher. Microorganisms in organic soils not only mineralise more actively, but also contribute to the build up of stable soil organic matter. The amount of microbial biomass and decomposition is connected: at high microbial biomass levels, little light fraction material remains undercomposed and vice versa. Thus, nutrients are recycled faster and soil structure is improved.

— *Microbial carbon*: The total mass of microorganisms in organic systems is 20-40 percent higher than in the conventional system with manure and 60-85 percent higher than in the conventional system without manure. The ratio of microbial carbon to total soil organic carbon is higher in organic systems as compared to conventional systems. The difference is significant at 60 cm depth. Organic management promotes microbial carbon and thus, soil carbon sequestration potential.

— *Enzymes*: Microbes have activities with important functions in the soil system: soil enzymes indicate these functions. The total activity of microorganisms can be estimated by measuring the activity of a living cell-associated enzyme such as dehydrogenase. This enzyme plays a major role in the respiratory pathway. Proteases in soil, where most organic N is protein, cleave protein compounds. Phosphatases cleave organic phosphorus compounds and thus provide a link between the plant and the stock of organic phosphorus in the soil. Enzyme activity in organic soils is markedly higher than in conventional soils. Microbial biomass and enzyme activities are closely related to soil acidity and soil organic matter content.

— *Wild flora*: Large organic fields (over 15 ha) feature flora as six times more abundant than conventional fields, including endangered varieties. In organic grassland, the average number of herb species was found to be 25 percent more than in conventional grassland, including some species in decline. Vegetation structure and plant communities in organic grassland are more even and more typical for a specific site than in conventionally managed systems. In particular, field margin strips of organic farms and semi-natural habitats conserve weed species listed as endangered or at risk of extinction. Animal grazing behaviour or routing activity was found important in enhancing plant species composition. Weeds influence the diversity and abundance of arthropods and flowering weeds are particularly beneficial to pollinators and parasitoids.

— *High-energy efficiency*: Organic agriculture follows the ecosystem theory of a closed (or semi-closed) nutrient cycle on the farm. Organic land management allows the development of a relatively rich weed-flora as compared to conventional systems. Some "accompanying plants" of a crop are considered useful in organic management. The presence of versatile flora attracts beneficial herbivores and other organisms that improve the nourishment of predatory arthropods. When comparing diversity and the demand of energy for microbial maintenance, it becomes evident that diverse populations need less energy per unit biomass. A diverse microbial population, as present in the organic field plots, diverts a greater part of the available carbon to microbial growth. This increases the turnover of organic matter, with a faster mineralization and delivery of plant nutrients. Finally, more organic matter is diverted to build up stable soil humus.

— *Erosion control*: Organic soil management improves soil structure by increasing soil activity and thus, reduces erosion risk. Organic matter has a positive effect on the development and stability of soil structure. Silty and loamy soils profit from organic matter by an enhanced aggregate structure. Organic matter is adsorbed to the charged surfaces of clay minerals. The negative charge decreases with increasing particle size. Silt is very susceptible to erosion since it is not charged, but organic-matter layers on the silt surface also favour aggregates with silt.

Genetic Resources for Food and Agriculture

There are several hundred millions of small farmers in the world who do not have the economic means to buy high yielding seeds or the synthetic fertilizers and pesticides necessary for conventional cultivation. Many of these have opted for the maintenance or reintroduction of organic systems based on traditional forms of agriculture.

These promote the use of varieties and breeds that are better adapted to local stress conditions and do not require unavailable inputs such as, for example, veterinary drugs. There are also farmers who have opted for organic agriculture, in part because they wish to produce healthy and environmentally-friendly food, and also because they are attracted by the strong demand for organic products and the related premium prices.

Market driven farmers should, as a minimum, rotate crops as the first step towards improving agricultural biodiversity. This is one of the methods required by organic certification bodies as well as by financial programmes. These farmers have also opted for sowing locally-adapted species and varieties that are more resistant to disease and local environmental conditions because synthetic fertilisers and pesticides cannot be relied upon.

The adoption of organic agriculture methods requires farmers to follow a series of agronomic practices that make organically managed systems biologically much more complex than conventionally managed systems. Organic farms make use of larger numbers of plant and animal species than conventional systems. As a result, the large pool of genetic resources for food is maintained and other useful organisms, such as predators, pollinators and soil microorganisms are increased—for the very benefit of the agricultural system.

The contribution of organic agriculture to the survival of centres of diversity and to underutilised species, varieties and breeds is described below. The specific requirements of organic farmers for a productive genetic material, in conditions of low-input and harsh environments, are generating innovative approaches to research and development; the emergence of participatory systems of selection and distribution of appropriate genetic material is presented.

The continued cultivation of crop species within their centres of diversity plays a fundamental role in the maintenance of genetic diversity. Preserving the integrity of centres of diversity through ecologically-sound agriculture is an indispensable inheritance for agriculture and as such, for food security for future generations. It is the genetic variability that allows populations to adapt to changing environmental conditions.

In centres of diversity, the introduction of organic practices is aiding the conservation, through cultivation, of populations with high genetic variability. The maintenance of agricultural production in centres of diversity requires market outlets for peasants and indigenous communities. The new income opportunities offered by organic markets reverse the present trend of abandoning land which has previously been economically uncompetitive.

Examples where organic agriculture created viable means for in situ conservation and use of areas with diverse genetic heritage include: producing and processing cocoa in Mexico to sell Maya chocolate to tourists, and the maintenance of naturally-pigmented cotton in Peru, resistant to pests, diseases and drought. These organic market opportunities provide for the economic viability of thousands of farmers and processors and maintain genetic variability for future generations.

In the past, agriculture has played an important role in the maintenance of genetic diversity. The substitution of a large quantity of species for only a few and the adoption of high yielding and uniform varieties from a genetic point of view, has caused a significant reduction in the genetic inheritance of cultivated species. Many agricultural species, varieties and breeds which have played an important role in the human diet and traditional cultures have practically disappeared over the last century. Organic farmers breed varieties for quality, nutrition, resistance and yield, in reduced input growing conditions. Research has shown that these characteristics are more likely to be found in older native cultivars.

In particular, open pollinated varieties and indigenous breeds offer diverse and regionally adapted characteristics that are better suited to organic agriculture. In the last decade, the adoption of organic agriculture has indirectly established a rescue process of species, varieties and breeds threatened by under-use or extinction. The restoration and enhancement of under-utilised species and varieties has been motivated by specific demands of both consumers and farmers.

For the rescue of varieties threatened by extinction, the development of a market is fundamental and it is here that organic agriculture plays an important role as the price premium gives an additional value to the product. This is especially the case now that there is a consumer' interest in traditional, speciality and organic products. Many consumers search for quality aspects, for either health reasons or culinary traditions.

Examples where organic agriculture has allowed the maintenance and improvement of species and varieties that otherwise would suffer strong genetic erosion or extinction include: the discovery of the nutritional value of the gluten-free quinoa in Peru and saraceno grain in Italy; re-introduction of local rice varieties in traditional diets and culture in Indonesia; and economic viability of the Garfagnana spelt in Italy. These cases provided for the survival of poor communities in marginal areas and valorised endangered genetic resources.

The majority of crop varieties available on the commercial market are not suitable for organic cultivation methods as they have been selected for production dependent on irrigation and large quantities of synthetic fertilisers and pesticides. Many of these are hybrids and are not open-pollinated. In the last few years, the problem has worsened following the arrival on the market of genetically modified varieties. The selection objectives of organic agriculture differ from those for conventional agriculture. It is of crucial importance to utilise the genotype potential for an improved adaptation of varieties to the low-input conditions prevailing in organic agriculture.

The necessity for organic farmers to find methods for obtaining quality products with good yields and limited production costs is greater than for other farmers. Besides the fact that organic farmers cannot apply synthetic inputs, their use of organic fertilisers, natural pesticides and other permitted substances is uneconomical in the long-term. Permitted external inputs as such are relied upon mainly during the conversion period to organic agriculture or under exceptional circumstances.

The comparative advantage of certain varieties to withstand local natural stress, especially in marginal areas, leads organic farmers to adopt biodiversity management as their main productive strategy. Empirical organic breeding systems have been based on the selection of individuals better adapted to the local environment and that are more resistant to pests and diseases.

Many of these systems have demonstrated interesting results in restoring and improving local varieties. Examples of restoration of varieties and breeds include: the rescue, in Germany, of an old variety of wheat with a vegetative cycle that allows the absorbency of nitrogen available in sandy soils; in Cuba, the success of local pumpkin varieties is used as the basis for the refinement of methodologies for the selection of varieties for low-input situations.

Animal breeding for high performance and selecting for early maturity have led to increased susceptibility to diseases, joint inflammation and mastitis as well as circulatory, metabolic and fertility problems of livestock. Loss of breeds is exacerbated by the narrowing genetic base of modern breeds and hybrid lines. The trend towards inbreeding increases the degree of genetic uniformity in the animals and hence, susceptibility to infection, parasites or epidemics. A significant proportion of local breeds remains in the care of pastoral people and traditional livestock owners in developing countries.

Local breeds are suitable for free ranging and robust, thus viable in marginal areas. While the yield may be less in the short-term, animals are more resilient and able to survive in the long-run. Examples where organic agriculture restored, through utilisation, genetic resources resilient to local natural stress include: productive rearing, in Italy, of authochthonous races of the Maremmana cattle, on the verge of extinction, due to its suitability to grow in marshy environments, and the re-establishment of native poultry in South Africa, due to their resistance to Newcastle disease.

Historically, farmers have managed many varieties and breeds according to agronomic and culinary properties. Considering the need for a wide gene pool to improve and multiply genetic resources for food and agriculture, breeding requires access to seeds and breeds from the formal and informal sectors. Open pollinated varieties, which represent an important gene pool for resource-poor farmers living in marginalised and stress-prone areas, are rapidly vanishing. They are replaced by very few hybrid varieties which require inputs not available to poor farmers and which entail dependence on large seed companies.

Limitations and threats associated with crops have stimulated many organic farmers, especially in the horticulture sector, to produce their own seeds. In order to do this, they have often had to rescue local varieties and develop their own system of selection and distribution. In many cases, the systems include the exchange of seeds between farmers as a fundamental instrument. Organic systems encourage the preservation and expansion of older, locally bred and indigenous varieties and breeds. Farmers who save their own seeds can gradually increase crop resistance to pests and diseases by breeding for "horizontal resistance".

Horizontal resistance is the ability of a crop to resist many or all strains of a particular pest. By exposing a population of plants to a certain disease or pest, then selecting a group of the most resistant plants and interbreeding them for several generations, a given population becomes more resistant than the original population. Horizontally resistant cultivars are well adapted to the environment in which they were bred, but may be less suitable for other growing conditions.

Benefits derived from new varieties bred by farmers require a legal system of common ownership that allows equitable access and benefit sharing. The biodynamic network of farmers and breeders in Germany provides an example of how such a system could be organised: trials, selection and evaluation of genotypes adapted to low-input conditions is made by farmers and common ownership of new varieties is shared by the community.

Organic agriculture is providing an important contribution to the in situ conservation, restoration and maintenance of agricultural biodiversity. The spontaneous establishment of participatory systems of research and development is shaping a simple and practical system of equitable sharing of benefits derived from genetic resources for food and agriculture. The growth pattern shown by the conversion to organic agriculture throughout the world suggests that this contribution is likely to increase still further.

WILDLIFE BIODIVERSITY

According to the IUCN Red List of 2000, approximately 70 percent of all endangered species of birds and 49 percent of all plant species are spoiled by agricultural activities and approximately 25 percent of the world's wild animals and plants is running the risk of extinction by the middle of this century. Agricultural productivity depends upon the maintenance of ecological balances and the natural properties of plants and animals.

The fundamental role of maintaining biodiversity is demonstrated through ecological services such as pollination of crops, predation for biological control of pests, microorganisms' maintenance of soil fertility and other services vital to the food web. On the other hand, agriculture has the same important role in wildlife conservation, provided that it avoids the use of substances that could have a harmful effect on natural species and that it maintains food and shelter through a diversified landscape and permanent vegetation.

Finally, a type of land use that provides suitable biological corridors is essential for wildlife conservation.

Nature conservation has traditionally consisted of geographically targeted efforts. While this approach has resulted in a number of successes for rare species or key locations, worrying declines of protected species have occurred. A healthy environment is a prime objective for the conservation of vital terrestrial ecosystems and the wildlife in it. Natural faunal and floral species have strong connections with agriculture, whatever their habitats are, especially as agricultural fields occupy much of the earth's land surface.

Protected areas simply cannot be viewed in isolation from the communities within and near them. Almost everywhere, rural dwellers claim historical rights on protected areas which governments have, at a point in time, declared "protected" for national interest. People inhabiting within or in the neighbourhood of protected areas depend directly on their resources for a living.

In India, for example, at least three million people live in protected areas and many other millions live in their proximity. In Latin America, about 86 percent of national parks are inhabited by indigenous people and migrants. This very dependency on protected areas and its diversity of life forms imposes ecological farming policies. If nature is to be protected successfully, protected area dwellers should be given agricultural choices which are not environmentally destructive and economically rewarding. Considering that the relationship between wild biodiversity and agriculture is reciprocal, the protection of wildlife, biodiversity and natural areas must include a correct management of agricultural systems.

Organic Agriculture and Nature Conservation

There is no doubt that farmers are the most important managers of natural resources. Several studies indicate that organic agriculture safeguards non-agricultural biodiversity and offers a viable alternative in protected area categories where human activities are allowed. Most importantly, the huge land surface surrounding protected areas requires an agro-ecosystem management that preserves the safety and integrity of the landscape.

If farm land bordering and connecting protected areas employ organic methods, there is no reason to fear the loss of wildlife or contamination of air, water and soil. These buffer zones are critical to the success of conservation in the protected areas. Organic agriculture enhances people's ability to live in harmony with nature and to derive economic benefit from their land.

Considering that most protected areas traditionally belonged to local villagers, organic agriculture allows local people to maintain some control over their land, protect land and biodiversity through their farming practices, reap its benefits for themselves and, at the same time, conserve and improve the natural environment. The direct impact that organic agriculture has on ecosystems can be seen on different scales: on-farm, farm margins, and overall ecosystem.

Organic Agriculture in Protected Areas

Certain protected area categories allow sustainable land-use activities such as organic agriculture, management of non-timber forest products, fishing, subsistence hunting and ecotourism. Organic farming within protected areas is a growing practice that can "help to define and control sustainable land uses in those protected areas containing significant human populations".

There are some 350 Biosphere Reserves in 85 countries to protect ecosystems. The area surrounding a biosphere reserve, named buffer zone, plays a critical role because activities carried out here strongly influence the core of the protected area itself. Often buffer zones are areas dedicated to agricultural practices. Conversion to organic systems can reduce the detrimental effects of conventional farming, and can provide sustainable systems, suitable for the management restrictions governing buffer zones, and consequently to natural ecosystem conservation. The protection of the natural heritage must consider the impact, be it positive or negative, that human activities have on it.

The characteristics of agriculture make it one of the main activities to be practised in protected areas and buffer zones. Organic agriculture offers a suitable alternative in ecosystems where geographic and morphological conditions are favourable to human activities, such as wetlands and lowland forests. Also, the ecological services offered by organic agriculture in biological corridors are of extreme importance.

Wetlands conservation

Wetlands are defined in the Ramsar Convention as "areas where water is the primary factor controlling the environment and the associated plant and animal life. They occur where the water table is at or near the surface of the land, or where the land is covered by shallow water". Wetlands are present in every country, from the tundra to the tropics. Their ecological importance derives from their capacity to host high concentrations of birds, mammals, reptiles, amphibians, fish and invertebrate species.

In fact, wetlands are boundary areas that combine the components of marine, fluvial and terrestrial ecosystems. In addition, many functions of wetlands derive from the interactions of different properties of soils, water, plants and animals. Some of these functions include, for example, water purification, water storage, flood mitigation, recharge and discharge of underground aquifers, and stabilisation of local climatic conditions. Natural wetlands are among the most threatened ecosystems in the world. Their high productivity and the morphologic characteristics make wetlands excellent areas for many human activities, especially agriculture.

Often the damage caused by land reclamation for agriculture or unsustainable practices leads to the disappearance of wetland areas and corresponding biodiversity. Suitable management is therefore a critical priority to save these fragile ecosystems. Organic agriculture can help wetland conservation, by providing suitable habitats for wildlife species, reducing water pollution and, at the same time, offering a valid economic alternative to the exploitation of natural resources.

Examples where the avoidance of synthetic inputs and cropping strategies employed by organic farmers in wetlands offered breeding and feeding habitats for endangered wildlife include: return of cranes and storks in the cereal production in Muravia Park, Russia and of egret and heron in rice production in the El Ebro Delta, Spain. Organic beef production in the Pantanal Region, Brazil, created natural grasslands vital to wild mammal herbivores in an area considered the world's biggest ecological sanctuary and which previously suffered from deforestation of savannah and implementation of artificial pastures for beef

Protected forest areas

Human activities such as modern agriculture and grazing can be a serious threat to forest ecosystems and hence, for a large percentage of the world's flora and fauna. In fact, the practice of clearing the tree cover often interrupts the continuity of canopy, an essential characteristic for genetic and specific biodiversity flow. In many cases, the same agricultural fields are also a barrier to wildlife movement, and pollution from agrochemical abuse may have negative impacts on forests.

Organic management in forest areas has the potential to reduce the loss of biodiversity caused by these agricultural activities. The organic systems utilised in forest areas are shade and sun cultivations. In the first case, the plants are grown under a canopy that consists of original forest trees or selected plants. In the latter, there is no forest cover on the agricultural field. Although yields of full sun production are often higher than for shade cultivations, the latter provide timber, fuelwood and other fruits, and have a lower soil erosion rate.

Shade plantations have been shown to be highly beneficial to biodiversity conservation in tropical forest ecosystems, including millions of migratory birds and other animals and plants. For example, the Smithsonian Migratory Bird Centre's research has demonstrated that bird species are almost twice as large in shade coffee as in sun coffee. Shade coffee cultivation therefore offers important conservation opportunities as the structural profile of shade coffee farms is similar to natural forest, providing habitat suitable for resident birds and migrants. As a result many scientists and conservationists believe that shade coffee plantations are ideal zones for migratory birds.

The Northwest Shade Coffee Campaign has obtained statistical data on shade coffee as compared to sun coffee and has shown how shade cultivations are preferred by bird species, mid-size mammals and several species of ants, beetles, amphibians, epiphytes and others. Smithsonian Migratory Bird Centre's studies indicates that at least 180 species of birds live in Mexican shade coffee and cocoa fields, and that 90 percent fewer bird species live in Colombian sun coffee plantations than in shade coffee.

Researchers highlighted the capacity of the canopy cover to support secondary structures, which in turn support arthropods, amphibians and other living beings. Another relevant fact is that shade coffee hosts a large density and diversity of predators and parasitoids involved in the control of insect pests. Shade coffee offers an optimal habitat also for other tropical forest species, such as beetles, ants, wasps and spiders, and supports a high diversity of many vertebrate groups. Finally, areas of high ecological value located around coffee farms are protected.

Examples of the co-existence of shade organic cultivations and richness of biodiversity in forest ecosystems include: shade coffee in the buffer zones of El Trionfo Biosphere Reserve, Mexico; shade cacao in the tropical forest of Montes Azules Biosphere Reserve, Mexico; and yerba mate in the threatened Atlantic rainforest of the Guayaki Biological Reserve, Paraguay.

Examples of conversion to organic agriculture to reduce pressure on endangered forests include: land exchange between Del Oro orange production and Guanacaste Conservation Area in Costa Rica to restore rare native forests; organic bananas to protect the threatened Guaraquecaba Atlantic forest and its extraordinary biodiversity in Brazil; and the reintroduction of native species through agroforestry in Ampay Forest Sanctuary in Peru. In all these cases, local communities' income was raised while providing benefits to the environment.

Biological corridors

Corridor zones link protected areas with one another and either remains under wild cover, or are managed to ensure that human land use is compatible with the maintenance of a high degree of biological connectivity. These areas have a fundamental ecosystem function forming part of a large-scale ecological web. To be effective, biological corridors must offer suitable habitats to wildlife, thus the maintenance of a healthy environment is necessary.

Areas joining parks or reserves are equally important for biodiversity conservation as they ensure ecosystem connectivity. These "linkage" areas often host agricultural systems but if they are managed in an unsustainable manner, the vitality of these corridors decreases or is totally hindered. Careful organic management in agricultural land between the protected areas has permitted the creation of important biological corridors and allowed the protection and increase of biological diversity.

Through on-farm structured vegetation and canopy tree diversity, organic agriculture has demonstrated its capacity to provide livelihoods to farmers while providing easier movement of animals between managed forests and protected areas. One of the largest, most unique conservation efforts presently underway in the world is the Meso-American Biological Corridor that runs across seven countries, covering an area of 770 000 km^2, from Panama to Mexico. Examples of organic cultivations include: shade coffee in the buffer zone between El Imposible and Los Volcanes National Parks in El Salvador where hundreds of endemic and threatened birds, mammals, reptiles and trees strive; and shade cacao and bananas in the Talamanca-Caribbean Biological Corridor and its buffer zones, Costa Rica, where most of the fauna is endemic.

Organic systems offer habitats similar to intact forests, providing suitable refuge for migratory and resident birds and foraging for many forest mammals, including species at risk. Organic land management and bird conservation programmes exist in several northern countries. Examples include: organic crop and livestock production in Brandenburg, Germany where the higher presence of skylarks has been observed; and organic farming inside a wildlife corridor that connects Adirondak National Park and Lake Champlain in the State of New York, USA.

REFERENCES

El-Hage Scialabba N., Grandi C. and Henatsch, C., "Organic Agriculture and Genetic Resources for Food and Agriculture," In *Biodiversity and the Ecosystem Approach in Agriculture*, Forestry and Fisheries, FAO, 2003.

Imhoff D., "Farming with the Wild: Agriculture and the Biodiversity Crisis. Biodiversity," *The Quarterly Newsletter of the CGBD*, 2001, Volume 11, Number 1.

Miller K., Chang E. and Johnson N., *Defining Common Ground for the Mesoamerican Biological Corridor*, World Resources Institute, 2001.

10

Promotion of Organic Farming

Organic Farming is becoming important in the agriculture sector in India, largely through the efforts of small groups of farmers. It has come out of the exploitative agriculture that has been followed by in all these years, resulting into damaging impacts on environment, human and animal health, soil and water resources. It is well known now that increased use of chemical pesticides and fertilisers have created chain of problems of soil, environment and water degradation.

The intensive chemical agriculture that has been followed after green revolution successes is causing heavy pollution of our food, drinking water, air, the life expectancy has improved, but the quality of life has substantially deteriorated. The rural economy is in ruins because of overdependence of outside inputs in agriculture such as seed, fertilisers, pesticides, growth-promoting chemicals etc. It is even said that the chemical agriculture has destroyed our ability to think about the right way to go forward.

Fortunately, alternatives to chemical agriculture are available in organic, biodynamic and eco-technological farming approaches. Though a small percentage of farmers are expected to take up organic farming, consumer demand for organically produced food and fibre products provide new market opportunities for farmers and farm-business around the world. In fact, Government of India has been clearly aware of the importance of organic and biodynamic farming approaches and the Ninth Five year plan document laid emphasis on 'Environment and sustainable agriculture', promotion of organically produced commodities, particularly in plantation crops, spices and condiments.

The Plan document emphasised the use of biofertilisers, bio-control agents, organic manures with infrastructural support. It is interesting to note that Food & Agricultural Organisation (FAO) Committee on Agriculture, during the 15th Session, has discussed the topic, "Organic Agriculture" and concluded that FAO has the responsibility to give organic agriculture a legitimate place within sustainable agriculture programmes.

In several developed countries, organic agriculture has come to represent a significant portion of food system and many other countries such as Japan, Singapore, France, United State of America etc. are experiencing growth rates that exceed 20 percent annually. Many developing countries began to seize the lucrative export opportunities presented by organic agriculture.

Realising the importance of organic and biodynamic farming, the Planning Commission, Government of India has constituted a Working Group on the subject. The Terms of reference of the working Group are:

— To review the performance of various programmes of Department of Agriculture and Co-operation (DAC) and ICAR undertaken on organic and biodynamic farming.

— To assess the technical soundness of organic and biodynamic farming practices to provide balanced nutrition and their efficiency to exploit the full genetic potential of the recommended crop varieties.

— To assess the techno-economic feasibility to such practices and their potentials and limitations to increase crop productivity and sustain food security of the country.

— To suggest measures/programmes for encouraging organic and biodynamic farming practices for which these are considered feasible and viable.

In consultation with Chairman of the Working Group, all members and some of the coopted members were contacted by providing them with available literature in the form of bulletins, books etc. and were asked to provide inputs for preparing this draft paper. M/S T.G.Kutty Menon, W.R.Deshpande, V.N.Shroff, D.V.Ragnekar, Labangalatika Dasi, Kapil Shah etc. immediately responded by providing their learned inputs, which have been taken into consideration in preparation of the paper.

Personal discussions of the Member Secretary with Shri Kuwarji Bhai Jadhav at New Delhi on 7th August provided some additional information, since he is Chairman of the Committee for the promotion of organic farming has been engaged in preparing a report for the Ministry of Agriculture, Government of India. In view of the shortage of time, the group could not meet to discuss the issues. The draft paper thus presented before the Steering Group on Agriculture and Allied Sections (constituted by the Planning Commission) on August 21, 2001. The suggestions of the members are also incorporated in this Final Report.

DEFINITIONS AND VARIANTS

Organic farming is basically a holistic management system, which promotes and improves the health of the agro-ecosystem related to biodiversity, nutrient biocycles, soil microbial and biochemical activities. Organic and biodynamic farming emphasises management practices involving substantial use of organic manures, green manuring, organic pest management pratices and so on. It has also come to mean that it is a system of farming that prohibits the use of artificial fertilisers and synthetic pesticides.

Biodynamic farming is an alternative variant where the chemical fertilisers are totally replaced by microbial nutrient givers such as bacteria, algae, fungi, mycorhiza, actinomycetes. Biological Pest management of crops is undertaken by employing predators, parasites and other plethora of natural enemies of pests, in addition to all the rest of option that help to avoid resorting to chemical pesticides. These agents could be augmented into farms or promoted through such activities that favour their flourished activities.

Composting, Green manuring, crop rotations,. Intercropping, mixed cropping etc. as well as bird perches, trap crops promote such biological activities. Although interpreted differently by various proponents, organic farming, biological farming, regenerative farming, biodynamic farming. Low external input sustainable agriculture (LEISA), low input sustainable agriculture (LISA) and sustainable agriculture connote the same ideology that provide integrated efforts to maintain agroecosystems with futurism.

Organic farming could then signify all such farming practices. Shri Kapil Shah indicated that a Organic farming is a philosophy for sustainable rural development and a not a technological or market option alone. This is a very important because organic farming cannot be viewed in isolation as for export market, but also could be a way of life in the Indian rural context.

Therefore, organic farming can be defined as socially just, environment friendly and economically viable alternative to chemical-oriented farming. An alternative system is called 'Eco-technological farming' has often been equated to organic and biodynamic farming in the Indian context. This system is an effective bend to traditional practices of wisdom with appropriate modern advances of science.

Integrated nutrient management (INM), integrated pest management (IPM) with optimum use of inorganic inputs is advocated. In fact, organic and biodynamic farming signify the production of a kind of chemical residue-free organic labelled products to sensitive consumers. In other words, Organic is a 'process claim' rather than a 'product claim'. This is because the agricultural products such as cereals, pulses, grams, fruits, spices or cotton cannot be distinguished into organic type or inorganic type.

Several countries and multitude of private certification organisations have defined in similar ways organic farming as agriculture. There were wide differences initially that the demand for constituency by multinational traders has led to great uniformity. Internationally networked non-governmental organisation, called International Federation of Organic Agriculture Movement (IFOAM), has permitted the use of certain products for soil conditioning and for pest and crop growth management. The basic rates of organic production are that natural inputs are approved and synthetic inputs are prohibited).

The Hanoi declaration of IFOAM emphasises that the Asian history of agriculture spanning into thousands of years, is in deep connection with cultural and ecological diversity. The erstwhile colonial rule as well as misdirected policies have undermined this balance. The increased food production in certain countries also paid for the cost of degradation of traditional diversity of crops and animals. Other international movements are Commission on Sustainable Development (CSD), Society for International Development (SID), International Union for Conservation of Nature (IUCN) and Bread for the world.

The Codex Committee of World Health Organisation on Food Labelling has been guiding the approval of such products. The broad definition of organic agriculture recommended by Codex Alimentarius Commission is more useful for practical purposes under Indian situation. "Organic agriculture is a holistic production management system which promotes and enhances agro-ecosystem health, including biodiversity, biological cycles and soil biological activity. It emphasised the use of management practices in preference to the use of 'off-farm' inputs, taking into account the regional conditions require locally adapted systems. This is achieved by using where possible agronomic, biological and mechanical methods, as opposed to using any synthetic materials to fulfil any specific function within the system.

Promotion of Organic Farming by ICAR

The seeds of commercial Indian "Organic cotton" cultivation were sown for the first time in Maharashtra in the early 1990s. Some progressive farmers, distressed by the negative effects of pesticides for insect suppression in cotton crop, reduced the chemical inputs and increased the use of organic manure, developed their own techniques to optimise resources in order to develop sustainable farm.

The pioneers, in this field, are M.V. Wankhede, S.P. Wankhede, R.S. Wankhede of Amaravati dist., Anantrao Subhedar, Om Prakash Mor and Tukaram Bhimsingh Jadhav of Yavatmal dist., or Shri Jain of Karanja-Lad who has a historical cultivation background of a few decades. They tested the 'Fukuoka' principles of farming, and stabilised their farms due to their ingenuous approaches. A team of CICR scientists visited the Yavatmal farms in 1992 crop season to analyse their package of practices.

Similar efforts of promotion of organic farming have been made in many states. Efforts have been made by the NGOs to study organic farming in Gujarat, Madhya Pradesh, Kerala, Karnataka, Tamil Nadu. Agricultural Universities organised workshops, Group meetings.

Seminars and Conferences on this topic drew attention of scientists to the need of research in this area. The use of biofertilisers, biopestides, vermicompost, Farmyard manure, green manure, crop residues have been based on long experimentation.

In fact, a number of farmers, NGOs and even some Universities/Institutions are practising organic farming, using traditional sources and methods of nutrient supplies to the crops and nonchemical forms of plant protection measures with varying degree of success. However, the technology adopted and methods followed are not well-documented. There has been a good suggestion from Shri Kapil Shah who emphatically proposed that the Universities and institutions that are undertaking organic farming trials should not only go about INM and IPM.

It is essential that the philosophy of organic farming are percolated to them properly. It is for the scientists, basically to evolve separate protocols for the conduct of organic farming trials. A distinction is necessary at this stage between the trials of organic farming and biodynamic farming including the eco-technological approach wherein the organic inputs approved by the certifying agencies are tested in replicated or large-plot trials instead of combining INM and IPM.

Due to indiscriminate use of hazardous insecticides for controlling cotton pests, the resistance of insects against the insecticide hiked up and in turn compelled the use of more number of sprays, and thus a vicious cycle is created, escalating cost of cultivation. Organic farming will help in reversing this trend. Evidences on poorer choice of multiplication rate of pests on organically grown cotton are encouraging factors to pursue this protocol. Scientists of Central Institute for Cotton Research, Nagpur had begun research in 1988 for identifying optimal practices for conserving soil moisture and also for improving organic matter content of marginal soils.

The results of these experiments led to the thought of comparing the three options, viz., inorganic, organic and an even mixture of both these in the third. It was proposed to begin a long-term demonstration trial at CICR on organic cotton cultivation and with comparative plots of inorganic and an integrated crop management plot, having 50% each of organic and inorganic components of inputs.

Improvement of marginal soils where cotton is extensively cultivated in Maharashtra, Madhya Pradesh and some parts of Andhra Pradesh, through certain methods such as in situ live green mulching, green manuring and addition of compost and recycled farm waste were in progress in the late eighties at Central Institute For Cotton Research. The continuous crop failure of farmers in the early '90s in spite of cultivation of hybrids was analysed to be due to poor expression of heterotic potential in marginal lands under rainfed conditions.

Associated techniques of farm waste recycling including cotton stalks using Trichoderma viride, the cellulolytic fungus were perfected. The sharp rise in cost of cultivation due to extensive applications of pesticides for hybrid cotton cultivars was found to be due to improper understanding of pest dynamics in cotton cultivation under high pesticide umbrella, which was the recommendation in vogue. The institute was involved in the search for alternate techniques of pest suppression in cotton such as by the use of bio-agents, pesticides of botanical origin etc.

During these field experiments as well an on-farm demonstrations, it became clear that there is a good scope for the reduction of agro-chemicals in cotton without any deleterious effects on crop yield. An ad-hoc field experiment was begun in 1991 at CICR farm in 600 sq.m. plot to marry the agronomic techniques to reduce the fertiliser component while the pest management was exclusively through biological means. LRA 5166 was the variety that was invading central zone from southern zone and was chosen as the cultivar for the experiment, although it was known to be more susceptible to jassids.

Parallel plots of similar dimensions were kept with the recommended practice of fertilisers and insecticides. A third plots received 50% of the inputs of both the other plots. Another set of 3 plots was cultivated with G. arboreum (Desi) cotton variety, AKA 5 and later with AKA 8401. This was planned as a long-term trial. The breeders had criticised the selection of LRA 5166, but the scientists went ahead with that since we had a conviction that crop husbandry could modify the growth pattern of cotton cultivars, as was seen in some farmers' fields in Vidarbha.

The scientists working in CICR had, by then, equipped for the research necessary in optimising the organic inputs for effective crop management. It explicitly brings out the fact that after third year, the yield of organic plots, which did not receive fertilisers and insecticides produced as much cotton as that cultivated with them. Simultaneously, it reduced the bollworm damage.

The following crops are currently cultivated under organic farming methods in our country. Cereals: wheat, paddy, jowar, bajra, maize Pulses: Pigeonpea, chickpea, greengram, blackgram, chana, Oilseeds: groundnut, Castor, mustard, sesame Commodities: Cotton, sugarcane, particularly for Sugarcandy (gur) Spices: Ginger, Turmeric, Chillies, cumin Plantation Crops: Tea, Coffee, Cardamom Fruits : banana, sapota, custard apple and papaya Vegetables: Tomato, brinjal, cucurbits, cole crops, leafy vegetables.

The initiatives taken by Bombay Burmah Trading Corporation (BBTC) to convert Singampatti group of estates in southern India to market organic tea internationally is the first of the few efforts for commercial organic agriculture. A number of organisations such as The Ecological Development Society in Pondicherry, Institute for Integrated Rural Development at Aurangabad, The Society for Equitable Voluntary Actions (SEVA) in West Bengal, The Indian Agency for Organic Agriculture (IAOA), Peekay Tree Crops Development Foundation (PTCDF) at Cochin undertake training of personnel towards organic farming. All India Federation of Organic Farmers promotes organic farming in the country.

It has been realised that India can play a major role in International organic spices market. The Spices Board has emphasised the expansion of Indian spice market because it is felt that the spices growers have a natural advantage in terms of large tracts of land in the tribal belts of Orissa, northeastern states, Nilgiri hills and Andaman and Nicobar islands where traditional practices are still in vogue.

The Board has already launched the Schemes to assist the organic spice growers by publishing national standards for organic spice production. It has received approval from the National Standards Committee of IFOAM. During 1997-98, India has exported 32.01 tonnes of different organic spices as against 25.32 tonnes in previous years.

The exports are expected to rise in the coming years as the Spice Board is assisting overseas buyers in soursing organic spices from India. According to Shri Kapil Shah, organic farming lead to the consider a good national food security option of healthy food for teaming population instead of creating an export-oriented business alone. In fact, it suggetsed that ICAR should take the lead for development of standards fore organic cultivation rather than APEDA of Commerce Ministry.

During the early part of last decade, the concept of Sustainable Agriculture and Rural Development (SARD) was introduced in one FAO conference held in Netherlands to attain food security, employment and income generation in villages and natural resource conservation leading to environmental protection. The intense development and progress of organic farming in our country also synchronised with this resolution as well as similar resolve of United Nations Development Programme.

Besides these efforts of ICAR and other NGOs in promoting organic and biodynamic farming, a real government support either in the form of subsidising the organic inputs or production promotion schemes of FYM, NADEP and other biological composting methods, AMRUTPANI, green manuring, recycling of farm wastes using earthworms or Trichoderma spp., botanical pesticides, biocontrol agents etc. has not been visible. There are sporadic efforts on the part of the State governments to organic and biodynamic farming.

R& D Issues

Examine the current economics of organic crop production system in the light of institutional support and based on the Chapter 10 agenda 21 of the Action Programme on Environment and development in Rio de Janeiro conference. On-farm research on standardised protocols for feasible cropping systems and their package of practices. Study and assess the likely environmental and economic values and externalities associated with the above conversion and its implications for future incentives to organic farming in the country.

Comparative evaluation of social and environmental economics are very much require to study both chemical and organic agriculture. While studying this, it is necessary to include all indirect costs including subsidiaries for the comparative analysis. Shri Jagdish Nazareth has given a very good suggestion on the nutritional security. It is now evident that food grains in the hands of government agencies at the central level are no insurance against hunger deaths, amongst landless labourer and tribal sections of rural community. There have to be an alternative approach to deal with this problem, according to him.

In fact, the centrally procured and stored food grain are not fit for exports, as evidenced by the recent rejection of 90,000 tonnes of wheat exported to Iraq under UN supervision. It is in this context, he suggests that there should be some crops and lands reserved for organic production only. Currently, medicinal plants, many of which are exported, are being cultivated using banned pesticides, on the basis of recommendations of the agricultural universities.

Pesticide residues in milk and milk products exceed the maximum permissible limits as specified by the Codex alimentarius. These are legal and not safety limits. It is therefore necessary to review all the pesticide recommendations made by agricultural universities and institutions. Pesticides pose the greatest hazard to our international export. In the light of their renewed resistance to pests, as well as the enhanced dosages applied by the users, the scientific recommendations of pesticide application for the control pests, diseases and weeds needs a critical review by an expert team.

Developmental Strategies

Various organs under different Ministries such as Commerce, Textiles and Agriculture should have an inter-ministerial group to promote the organic crop production and business interests. For example, the Agricultural Processed Products Export Promotion Development Agency [APPEDA] under Ministry of Commerce has brought out in March, 2000 the National Programme for organic Production containing the Standards for Organic Products. APPEDA is known to further refine this 48-page document and also authenticate all the current organic certifying agencies of the world, which have set up offices in India for being licensed for this job.

Organic farming brings earnings back to the rural area. In the western countries, most farm expenses are made to capitalistic factories to pay for machinery, fertiliser, pesticides and additives. In our small holdings we do not need heavy machinery, and if our manure and IPM agents are made on our farms, we keep our people employed and their wealth with themselves without transferring it to townsfolk. This may be bad for the government since as no cash is transacted no taxes can be levied. This economic insight should be properly developed with some case examples to show increase in rural wealth.

All sources of organic material that can be used as manure should be identified, this should include industrial wastes also. Gaps in technology that prevent the utilisation of some wastes should then be identified. This should be done to satisfy critics that not enough organic material is available for organic farming. Another intermediate farming practice that uses a stimulatory dose of fertilisers along with manures should be also be developed. As a lot of organic carbon and fixed nitrogen is lost in the process of composting, other methods of stabilising organic manure should also be looked into.

In today's climate of scepticism and lack of encouragement towards organic farming, and subsidies for fertiliser use, one needs to be a die-hard believer to practice organic farming. Such people tend to get carried away when talking about their beliefs and some loose statement is caught hold of by the opponents to ridicule a whole set of otherwise proper arguments. The analogy of manure as food and fertiliser as tonic should be developed and widely disseminated. It is something people find easy to understand and correlate the consequences.

Marketing Infrastructure

Strong linkages between growers and consumers with minimum influence of middlemen should be developed. If only, government infrastructure such as agriculture, horticulture, animal husbandry and fisheries are integrated, this could be successfully organised. Sensitisation the philosophy and ideals of organic farming as a way of life to these agencies through a 'bottom up' approach with suitable delegation of powers and authority would accelerate the facilitation of marketing in the country in the line of many such success stories that are paradigms for dairy and processed food in the last few decades.

Farmer friendly shanties, in the model of 'Ritu bazars of Andhra Pradesh could be opened through non-profit organisation or farmers' groups. Suitable marketing strategies and designs should be developed for the build up of native markets in all the states. Shri Kapil Shah has suggested that for certification of organic farms and their products, it is necessary that the regional level organic standards be first developed which could be then be coalesced to form the National standards.

In this regard, Shri L.N. Modi suggested that there is enough potential for exploring the organic products markets. Motivation of hotels, large restaurants, airlines, railways, departmental chain shops is becoming possible now with awareness of quality food. It is learnt that Swiss Air has switched over to organic meals to be served worldwide on all their flights. There are already departmental supermarkets in large cities of Delhi, Mumbai, Chennai and Bangalore where the organic agricultural produce are marketed in small scale and in the years to come, there is growing demand for organic food, not only outside India, but within India also.

Certification of organic farms, their produce and products along with providing logos and accreditation powers should be initiated by regional non-profit organisations, who could be empowered for this by the government with the condition that they will not commercialise these efforts and exploit the growers and consumers alike. FAO believes that it is important for developing countries such as India to certify organic food products so that they can compete in the extraordinarily growing International market.

The return from organic agriculture can potentially contribute to food security by increasing farm income. Shri Jagdish Nazreth is of the clear opinion that there are some vested interests that wish to use organic standards to erect non-tariff barrier against Indian exports.

The December 2000, Stockholm convention on POP, means that the most Indian agricultural exports will be affected and can because of pesticide residues that are commonly found in most Indian agricultural commodities. The government of Indian regulation banning uncertified organic agricultural products will mean that organic exports from India will face to sets of regulations and at least for some years, very little will pass through them for exports.

By following the lead set by some vested interests, perhaps, with the best of intentions, we will be shooting ourselves in the foot, with the complex armchair organic standards that are being devised in certain quarters. Therefore, he suggested that one should bring down the pesticide residues in general agriculture by organic methods. This will certainly work rather than trying to create unreachable standards for a niche market, which might never develop.

Government Support at Village Level in Farm Subsidies

The following items deserve attention of the Planning Commission for being recommended for government support in the Tenth five year plan.

— Manufacture of composts

— Manufacture of biofertilisers

— Manufacture of bio-pesticides

— Manufacture of small-farm implements

— Market intelligence and information system

Establish credit-linked and market-assured production units Development of certification agencies through ono-profit organisations Development of marketing systems through non-profit organisations and organic farmers' groups The major concern in organic farming, which has been raised time and again by agricultural scientists and planner alike, is regarding the timely non-availability of several such useful products that are permitted for inorganic agricultural products.

Hence, there is often a setback in the input chain and the farmers, even though desirous to adopt the organic recommendations, are deprived of the opportunity. Promoting the production of these inputs at village level can mitigate this situation. This would ensure generation of employment and income in rural areas. There are several small units in the country that are producing these inputs for organic and biodynamic farming. For example, several units for massproduction of bioagents such as friendly insects, pathogens etc. are operational. However, they are unorganised and scattered.

The products such as neemjeevan spray from the wonder-tree are also marketed. Several herbal products have been tested in the Agricultural Research system and have been reported to be efficient in management in pests, soil fertility etc. Smt. L. Dasi suggested that there should be a change for discontinuing the subsidy for export of leather and meat, enhancement of cattle heads to meet the hiked national needs of urine and dung for organic compost preparation. The government must stop all kinds of subsidies for export of leather and meat, which will help preserve the cattle population.

Simultaneously a massive programme for increasing healthy cattle population is necessary so that the organic compost inputs of urine and dung are freely available. Qualified subsides should be for indigenous bullocks for draught purposes and for preservation of native animal biodiversity. One of the significant promotion activities that has been suggested by Smt. Dasi is regarding the improvement of the fodder situation by rejuvenating the cultivation of common fodder species. The trend of butchering of the draught animals need to be arrested and their population need to be supported by proper availability of feed. This will encourage for more production of organic inputs such as urine and dung.

Shri Jagdish Nazareth has given very important suggestions. He suggests that the Green revolution model has limits in application to approximately 50 million hectares. Thus, the remaining 50 million hectares out of the officially 100 million hectares needs fresh approach. He also suggests that 40 million hectares of unofficial farm land that is degraded can be placed in the organic farming strategy. Drawing inspiration from the recent Cuban agriculture experience, it is wise to consider the production of compost in large scale through industrial plants utilising earthworms and other biological converters as well as by extensive adoption of green manuring of agricultural lands.

Cuba produced 93,000 tonnes of worm humus through 172 vermicompost centres, spread across the island way back in 1992. Each Indian State government could emulate this. Similarly, Cuba has also developed several centres for production of predatory ant through low-cost techniques for the management of various pests in sweet potato. Our country also could invest to patronise such efforts through NABARD and other such agencies that have been already sensitised in this regard by CICR.

REFERENCES

Dorothy Myers and Suer Stolon (ed.), *Organic cotton from field to final product, Intermediate Technology,* Publication, 1999, 250 p.

Louis C. Kervran, *Biological Transmutations,* Published by Swan House Publishing Company, Bingnampton, N.Y. 1972.

Pfeiller E., *Biodynamic Farming and Gardening,* Published by Pauma Valley Company, California, 1973.

Thampan P.K., *Organic Agriculture,* Peekay Tree Crops Development Foundation, Cochin, Kerala 354 p, 1995.

Tandon, H.L.S. (Ed.), "Fertilisers, organic manures, recyclable wastes and biofertilisers – components of integrated plant nutrients," *Fertiliser Development and Consultation Organisation,* New Delhi, 1994, pp.148+viii.

Bibliography

Alonso Villalon, N., "First Spanish aquaculture certification includes sturgeon", *The Organic Standard* 12: 14-15, 2002.

Altieri, Miguel., "Sustainable agriculture," In *Encyclopedia of agricultural science*, v. 4. Charles J. Arntzen and Ellen M. Ritter, (eds.), San Diego, Academic Press, c1994. p. 239-247.

Anon, *Green Revolution to Organic Revolution, Comprehensive Agric-Commodity Intelligence*, April, 2003, pp. 21-35.

Barret, J.R., "Livestock farming: eating up the environment?", *Environmental Health Perspectives*, 109(7): A312-A317, 2001.

Basu, A. K. and Tanweer, A., *J. Indian Soc. Cotton Improv.*, 2005, 30, 1–20.

Christensen, D.K., Jacobsen, E.M. & Nohr, H., "A comparative study of bird faunas in conventionally and organically farmed areas", *Dansk. Orn. Foren. Tidsskr* 90: 21-28, 1996.

Deb, DL and Joshi, HC, "Environment Pollution and Agriculture," in *Natural Resource Management for Sustainable Agriculture and Environment*, Angkor Publications (P) Ltd., New Delhi, 1994, pp. 227-293.

Dorothy Myers and Suer Stolon (ed.), *Organic cotton from field to final product, Intermediate Technology*, Publication, 1999, 250 p.

El-Hage Scialabba N., Grandi C. and Henatsch, C., "Organic Agriculture and Genetic Resources for Food and Agriculture," In *Biodiversity and the Ecosystem Approach in Agriculture*, Forestry and Fisheries, FAO, 2003.

Fowler, C. & M, P., *Shattering: Food, Politics, and the Loss of Genetic Diversity*, University of Arizona Press, Tuscon, 1990.

Geier, Bemwad, "International Federation of Organic Agriculture Movements," in *Sustainable Agriculture Solutions: The Action report of the Sustainable Agriculture Initiative*, Novello Press, London, 1999.

Imhoff D., "Farming with the Wild: Agriculture and the Biodiversity Crisis. Biodiversity," *The Quarterly Newsletter of the CGBD*, 2001, Volume 11, Number 1.

Klinkenborg, Verlyn., "A farming revolution: sustainable agriculture," *National geographic*, v. 188, Dec. 1995: 60-88.

Parrott N. and T. Marsden, *The Real Green Revolution – Organic and agroecological farming in the South*, Greenpeace, 2002.

Pfeiller E., *Biodynamic Farming and Gardening*, Published by Pauma Valley Company, California, 1973.

Prakash, TN, "A Theoretical Framework To Promote Organic Produce Marketing in India," *Indian Journal of Agriculture Marketing*, Conference Number Special, 2003, pp. 1-16.

Rao, PS, "Marketing of Organic Produce of Wheat in Rajasamand District of Rajasthan," *Indian Journal of Agriculture Marketing*, Conference Number Special, 2003.

Sanders, R., "A Market Road to Sustainable Agriculture? Ecological Agriculture, Green Food, and Organic Agriculture in China," *Development and Change* 37(1), 2006.

Scialabba, N. & Hattam, H., "Organic agriculture, Environment, and Food Security," *Environment and Natural Resources Series No. 4*, FAO, Rome, 2002.

Singh, GR, "Organic Farming for Sustainable Agriculture," *Indian Farming*, June, 2001, pp. 12-14.

Thampan P.K., *Organic Agriculture*, Peekay Tree Crops Development Foundation, Cochin, Kerala 354 p, 1995.

Veeresh, GK, "Organic Farming Ecologically Sound and Economically Sustainable," *Plant Horti Tech*, 1(3), Nov-Dec, 1999.

Index